A Monograph of *Phoenix* L. (*Palmae*: *Coryphoideae*)

Sasha C. Barrow[1]

Summary. Thirteen species are treated including one new species from the Andaman Islands, *P. andamanensis*, and two varieties within *P. loureiri*, var. *loureiri* and var. *humilis*. Species limits and distributions are defined, and aspects of morphology and lamina anatomy are examined in relation to ecology. Systematic analyses of the genus combine data from studies of morphology and lamina anatomy with DNA sequence data of the 5S spacer region (nuclear ribosomal DNA). The origin of *P. dactylifera* is discussed in the light of the results of the systematic analysis.

INTRODUCTION

The genus *Phoenix* L. (*Phoeniceae*: *Coryphoideae*) is a distinctive and well-known palm genus which, prior to this study, was regarded by Uhl & Dransfield (1987) as comprising 'approximately 17 species'. An Old World genus, *Phoenix* ranges from the Canary Islands through subtropical and tropical Africa, the Mediterranean, the Arabian Peninsular, the Indian Subcontinent and Indochina to Hong Kong. *Phoenix* species can be found in a range of different habitats from sandy scrubland at sea level to pine forest understorey at 2000 m, and from wet mangrove margins to semiarid zones. Despite a certain tolerance of some species to high salinity, atmospheric aridity and heat, all require constant moisture about the roots, and are often found in boggy or seasonally-flooded areas. In the driest environments *Phoenix* species act as good indicators of ground water; for example, the date palm, *Phoenix dactylifera*, is the well-known symbol of oases. Throughout the range of the genus, many species of *Phoenix* are used by man for food, clothing, construction, fibre and ornamental purposes.

MORPHOLOGY

Growth form

All but four species (*P. canariensis*, *P. rupicola*, *P. sylvestris* and *P. andamanensis*) produce basal suckers which may or may not develop into full-sized stems. In some species suckers remain at the stem base (e.g., *P. loureiri* and many cultivars of *P. dactylifera*). In other species (*P. paludosa*, *P. roebelenii*, *P. reclinata* and *P. theophrasti*) suckers develop vertically giving rise to clumps of equal-sized stems. The ability to cluster varies between species within *Phoenix*, but expression of this ability varies within species. It is probable that expression of the clustering potential is determined, in

Accepted for publication May 1998.
[1] The Herbarium, Royal Botanic Gardens, Kew, Richmond, Surrey, TW9 3AB, U.K.

part, by environmental factors. For example, expression of the clustering potential in the polymorphic species, *P. loureiri*, appears to be restricted to individuals under certain environmental pressures. From India eastwards to Indochina and the Far East, in marginal, often anthropogenic and fire-damaged areas, *P. loureiri* palms are predominantly short and often suckering. In contrast, individuals growing in undisturbed areas tend to be taller and solitary, though not exclusively so.

Leaf

The leaves of *Phoenix* are easily identifiable by virtue of four unique characteristics. Firstly, *Phoenix* is the only induplicately pinnate-leaved genus in the otherwise palmate-leaved *Coryphoideae*. Induplicately-pinnate leaves do occur elsewhere in the *Palmae* (in the *Caryoteae*) but they arise by a different mode of development. Secondly, the leaves of *Phoenix* are characterised by formation of a layer of tissue on the adaxial surface of unexpanded leaves. This tissue, known as the 'haut' on account of its skin-like appearance is formed by fusion and subsequent abscission of adaxial lamina ridges (Padmanabhan 1963; Periasamy 1967; Kaplan *et al.* 1982). It is not yet clear what function, if any, is associated with the haut. Thirdly, the leaflets of *Phoenix* have no true midrib but the central region is occupied by a band of expansion cells in which only minor vascular bundles and fibres are scattered.

Inflorescence

All species of *Phoenix* are dioecious [although DeMason & Tisserat (1980) reported the occurrence of apparently bisexual flowers in *P. dactylifera*]. Staminate and pistillate inflorescences are superficially similar and are much simpler in construction than those of most coryphoids. The prophyll is coriaceous, and splits either along or between margins to reveal a solitary inflorescence. The peduncle is flattened, bears no peduncular bracts and may elongate greatly with maturity to present the infructescence far beyond the prophyll. Staminate rachillae are of the first order and are flexuous, unbranched, and congested in their arrangement on the rachis, and bear flowers along their entire length. Pistillate rachillae are more robust, less numerous, very rarely branched to the second order, and are arranged in loose spirals or irregular clusters along the rachis. Elongation of pistillate rachillae with fruit maturation results in a general expansion and relaxation of the inflorescence that allows for more effective presentation of fruit for dispersal.

Floral morphology

Staminate and pistillate flowers of *Phoenix* are morphologically indistinguishable until late in development (DeMason *et al.* 1982) but are distinct at maturity. The basic floral structure in palms is trimerous, with a perianth of two slightly imbricate whorls each of three segments, six stamens in two whorls and three distinct uniovulate carpels (Uhl & Dransfield 1987). Flowers of both sexes of *Phoenix* differ from this basic palm flower structure only in corolla characteristics.

Staminate flowers are borne singly, each subtended by a small, non-persistent bract. They range in colour from pale yellow-white to yellow-brown. The flowers can be faintly sweet-scented, often turning pungently musty post-anthesis. As in the pistillate flowers, the three sepals are connate into a short cupule, the lobes variously distinct. The corolla consists of three petals, slightly valvate in bud, opening on maturity to expose the anthers. The congested arrangement of flowers along the rachillae prevents complete spread of petals and can cause distortion of flower shape. The six stamens are arranged in a ring. The anthers are yellow-white to yellow-brown in colour, linear-latrorse in shape and are borne on very short, erect filaments. Copious quantities of pollen are released on anthesis. Pollen grains of *Phoenix* are monosulcate and broadly symmetrical or slightly asymmetrical and elliptic in shape (in polar view). Ranging from $16 - 30 \times 9 - 15$ µm in size, pollen grains of the genus are amongst the smallest of all coryphoids (M. Harley, *pers. comm.*). The exine is tectate and reticulate, sometimes only finely so.

Pistillate flowers of *Phoenix* are more or less uniform across the genus. Each pistillate flower consists of a cupular calyx of three connate sepals, with variously distinct lobe apices, and a corolla of three imbricate petals. Of the three free carpels only one generally reaches maturity. The gynoecium is ringed by six minute staminodes.

Fruit

Fruits of *Phoenix* species are one-seeded berries with a smooth epicarp, variously fleshy mesocarp and silvery, membranous endocarp. Fruits mature through a range of colours from green to yellow to orange, ripening either golden yellow-orange (*P. sylvestris* and *P. canariensis*), bright orange (*P. reclinata*), greenish-brown (*P. roebelenii*), or various shades of red, brown, purplish-brown or black. In *P. dactylifera* the long history of cultivation and human selection has resulted in wide variation in date fruit colour, taste, texture, sweetness and size.

Seed

Seeds of *Phoenix* species are characterised by a deeply grooved raphe, of varying width and depth, running longitudinally along the seed. Seeds range in length from 7 mm (*P. roebelenii*) to 30 mm (cultivars of *P. dactylifera*). In transverse section, seeds are generally circular but those of *P. paludosa* are slightly flattened dorsiventrally. The endosperm of *P. andamanensis* is intruded by elaborate seed coat ruminations. Seeds of all other species have homogeneous endosperm with seed coat intrusion into the endosperm restricted to the region of the raphe. Embryo position is basal in *P. paludosa*, but more or less equatorial opposite the raphe in all other species. A small depression in the testa marks its exact position.

LAMINA ANATOMY

Materials and methods

Lamina sections were prepared for anatomical study following methods outlined in Martens & Uhl (1980). Cuticle preparations were made using a variation of the method of Alvin & Boulter (1974).

Description of Phoenix lamina anatomy

Lamina isolateral or dorsiventral, c. 175 – 400 mm in thickness. *Hairs* absent from lamina of most species, but abaxial surfaces of some species with scurfy, white ramenta (thin, membranous, flattened scales) along veins, or sclerotic plugs of tannin-filled cells. *Cuticle* of varying thickness, adaxial cuticle generally thicker than that of the abaxial surface. *Epidermis* variably cutinised or waxy. Epidermis of both surfaces of isolateral species and abaxial surface of dorsiventral species differentiated into costal bands of elongated, narrow cells and intercostal bands of more isodiametric cells; adaxial epidermis of dorsiventral species uniform with rectangular cell shapes. *Stomata* restricted to intercostal regions, solitary or grouped in regular or irregular files; equally abundant on both surfaces of isolateral species, very sparse adaxially in dorsiventral species. *Guard cells* slightly sunken or not, each with two conspicuous cutinized ledges, surrounded by four subsidiary cells (structurally specialised cells) in tetracytic arrangement. *Hypodermis* of one (rarely two) layer(s) beneath each epidermis, variously interrupted by stomata and fibres, occasionally sclerotic. *Fibres* abundant, vascular or non-vascular. *Non-vascular fibre bundles* adjacent to either hypodermis or epidermis or occasionally in mesophyll. Sub-hypodermal fibre bundles generally larger beneath adaxial surface than abaxial surface. *Vascular fibre bundles* forming sclerenchymatous inner bundle sheaths (IBS) of varying thickness, around vascular tissue. Stegmata (specialised cells containing silica bodies) associated with all fibre bundles. *Mesophyll* of thin-walled chlorenchyma cells. Cells elongated and palisade-like beneath both surfaces of isolateral species, beneath adaxial surface only in dorsiventral species. Central mesophyll cells uniformly isodiametric. *Idioblasts* containing raphides occasional in central mesophyll. *Veins* centrally positioned in lamina. Main vascular bundles (MVBs) interspersed with a varying number (most commonly a 'base number' of three) of small vascular bundles (SVBs). MVBs attached to each hypodermis by large cells without contents, which together form bundle sheath extensions (BSEs). Outer bundle sheath (OBS) consisting of colourless parenchyma cells, completely surrounding small veins, only partially surrounding MVBs. Transverse commissures (lateral veins) narrow, connecting longitudinal veins of all sizes. *Xylem* usually one large vessel within scattered parenchyma. Phloem vessels grouped in one strand. *Phloem* vessels often surrounded by a net of sclerotic cells. *Midrib* absent in *Phoenix*; instead, the central region (expansion zone) is occupied by a mass of expansion cells amongst which fibres may be scattered. Expansion cells in palms are colourless hypodermal cells which enlarge considerably, increasing their volume several times, to bring about unfolding and expansion of the lamina (Tomlinson 1990). Epidermal cells of expansion zone distinctly papillose. *Leaflet margin* acute or truncate in transverse section, with sclerotic scar tissue in the position of the primordial split. In some species (*P. paludosa* and *P. loureiri* var. *loureiri*) there is a proliferation of sclerotic, tannin-filled cells at the margin. *Tannin* deposits present to varying degrees in mesophyll cells, inner-bundle sheath fibre cell lumen, ramenta cells, in sclerotic cells at leaflet margin and expansion cells in midrib region.

Lamina anatomy and ecology

In any systematic study of palm lamina anatomy it is important to appreciate the effects of both lamina maturity and ecology on anatomical variation. Within this study all samples comprise a central portion of a median leaflet from a mature leaf.

No explicit attempt was made to survey how lamina anatomy changes during development. The determining role of ecological factors on characteristics of palm lamina anatomy has generally been neglected, but was discussed by Barfod (1988) and Zona (1990), and general observations of variation in *Phoenix* are given here.

Effects of ecological factors on *Phoenix* lamina anatomy are evident in characters such as cuticle thickness, degree of phloem sclerification, size and number of subhypodermal fibre bundles, number of MVBs and SVBs, and tannin distribution and quantity. Within *Phoenix*, cuticle thickness and degree of phloem sclerification appear to increase broadly with environmental aridity, as suggested by Tomlinson (1990). The species of *Phoenix* with thickest cuticles and most heavily sclerified phloem tissue are *P. caespitosa* and *P. dactylifera*, both species of arid lands. The correlation of cuticle thickness with xerophytic environments was reported elsewhere in the palms by Tomlinson (1990), Barfod (1988) and Zona (1990). The frequency and distribution of tannin cells varies widely both within and between *Phoenix* species. Only in cells of the leaflet margin is tannin presence of diagnostic value within *Phoenix*. Leaflet margins of *P. paludosa* are always marked by tannin-filled cells, visible to the naked eye. Similarly, leaflet margins of *P. loureiri* var. *loureiri* from Indochina and the Far East are almost always marked by tannin-filled cells. However, *P. loureiri* var. *humilis* palms of India and Nepal lack this tannin. The correlation of tannin presence with specific ecological factors is not clear but it appears that age and environmental or seasonal variation have a determining role and therefore tannin distribution is here considered to be of limited systematic use within *Phoenix*. In contrast, Zona (1990) found tannins to be highly informative in systematic analysis at the species level within *Sabal*, despite wide intraspecific variation.

Inclusion of ecologically adaptive characters into an analysis can be systematically informative, particularly at the species level (e.g., Zona 1990). Systematic analysis of such characters (suspected to be ecologically adaptive) in combination with data from other sources helps to identify those determined in part by ecology. It is only by analysis that the degree to which such characters are systematically informative can be appreciated. It is particularly important to consider ecological factors in a genus such as *Phoenix*, the species of which grow in habitats ranging from desert oases, mangrove margins, and steep limestone cliffs, to river margins.

This study has investigated the role that ecology may play in determining lamina symmetry. Tomlinson (1961) described symmetry of palm laminas in transverse section as being either isolateral (symmetrical) or dorsiventral (asymmetrical). Within the *Palmae* presence of an isolateral lamina is almost entirely restricted to subfamily *Coryphoideae*, where its presence appears correlated with ecological factors, particularly aridity. The tribe *Borasseae* provides a good example. Of the eight genera, *Borassus*, *Bismarckia*, *Hyphaene* and *Medemia* occur in dry, semiarid areas and all are isolateral. *Latania*, a genus of coastal cliffs and savannahs, was recorded by Tomlinson (1961) to be 'isolateral or somewhat dorsiventral', with stomata either equally abundant on both surfaces or more abundant abaxially. The remaining genera *Borassodendron*, *Lodoicea* (Tomlinson 1961) and *Satranala* (Dransfield & Beentje 1995) occur in wet areas and are dorsiventral. Variation in lamina symmetry is not restricted to the generic level, but is found also at species levels within *Phoenix*, *Sabal* and *Livistona*. Most *Phoenix* species are palms of semiarid regions and have isolateral laminas. The four dorsiventral species in the genus are *P. paludosa* (mangrove margin areas), *P. roebelenii* (on riverbanks and cliffs) and *P. andamanensis* and *P. rupicola* (warm wet forest on steep rocky hillsides, ravines and cliffs). *Phoenix reclinata* has the dorsiventral character of a clearly defined adaxial palisade layer but the isolateral character of equally abundant stomata on both lamina surfaces.

MOLECULAR STUDY

In a genus such as *Phoenix*, with a paucity of systematically useful gross morphological and anatomical characters, the advent of molecular techniques provides the prospect of exciting new data sets. For the purposes of a species level study of *Phoenix*, the *5S* intergenic spacer of the *5S* DNA unit was chosen for sequencing. In most eukaryotes, genes encoding *5S* RNA occur in long tandem arrays comprising up to several thousand copies, at (rarely) one to (more commonly) multiple chromosomal loci within the nuclear genome (Sastri *et al.* 1992). Each *5S* DNA unit comprises a coding gene of 120 base pairs long and a non-coding spacer region ranging in length from 100 – 700 base pairs (Sastri *et al.* 1992).

Materials

Fresh or silica-dried leaflet material of newly-expanded leaves were used for DNA extraction. All but two species of *Phoenix* and one species of *Sabal* were included in the molecular study. Where possible each species was represented by multiple samples. All samples are vouchered by herbarium collections deposited in herbaria in the country of origin and at RBG, Kew.

Methods

All DNA extractions were carried out according to the shortened cetyl-trimethyl-ammonium bromide (CTAB) method of Doyle & Doyle (1987), followed by a final purification using equilibrium density centrifugation in caesium chloride/ethidium bromide. The intergenic spacer region of *5S* DNA units was amplified by polymerase chain reaction (PCR) using the standard PCR reaction mixture (Saiki *et al.* 1988), using the universal primers *PIII* and *PIV* (Cox *et al.* 1992). Purified PCR products were sequenced in both directions (5' and 3') with the method of Sanger *et al.* (1977) using the *Taq* Dye Deoxy Terminator Cycle Sequencing kit and an ABI 373A DNA sequencer. Nucleotide sequences were initially edited and assembled using Biosystems programs Sequence Navigator MacApp and AutoAssembler MacApp version 3.0.1. Sequence alignments were initially produced with Clustal sequence alignment software (Higgins *et al.* 1992) and then improved by eye.

Results

5S spacer regions of 26 samples of 11 species of *Phoenix* and outgroup *Sabal mauritiiformis* were sequenced. *Phoenix andamanensis* was excluded due to lack of suitable material, and all attempts to obtain a *5S* spacer sequence of *P. acaulis* failed. *5S* spacer sequences of *Phoenix* range in length from 246 – 349 base pairs. The *5S* spacer region of *Phoenix roebelenii* is considerably longer than that of all other species in the genus due to presence of an indel (insertion/deletion event) 113 base pairs long. Alignment of 5S spacer sequences of species of *Phoenix* was problematic and there are areas in which alignment is suboptimal, even arbitrary. To maximise sequence alignment, gaps of one to three base pairs in length, were introduced. Even after introduction of gaps, sequence alignments remain far from satisfactory; nevertheless, all regions of the 5S spacer were included in the analysis.

SYSTEMATIC ANALYSIS

This study offers the first attempt at examining species relationships within *Phoenix*. By clarifying species limits and differentiation with *Phoenix*, a systematic analysis of the genus using cladistic methods becomes possible.

Data

Ten morphological and five anatomical characters scored for all species of *Phoenix* were included in the analyses. Character states and the data matrix are given in Tables 1 and 2. Molecular data, comprising 5S spacer sequences, were obtained for all but two species of *Phoenix*. Morphological, anatomical and molecular data were obtained for *Sabal mauritiiformis* which was included as outgroup.

TABLE 1. Morphological and anatomical characters for systematic analysis of species of *Phoenix*.

Character	Code
1 Habit	clustering 1, solitary 0,
2 Crown hemispherical	absent 0, present 1
3 Elongate internodes	absent 1, present 0
4 Persistent leaf base remains	absent 0, present 1
5 Leaflets to 3 cm wide, closely and regularly inserted in one plane	absent 0, present 1
6 Staminate petal apex acute to acuminate with jagged margins	absent 0, present 1
7 Pistillate flowers subtended by bractiform swellings	absent 0, present 1
8 Embryo position	lateral 0, basal 1
9 Endosperm	homogeneous 0, ruminate 1
10 Adaxial stomata	rare 0, abundant 1
11 Prominent adaxial palisade layer	absent 1, present 0
12 Non-vascular fibres in mesophyll	absent 0, present 1
13 Prominently persistent ramenta on abaxial surface	absent 1, present 0
14 Tannin-filled sclerotic cells along midrib and margin	absent 0, present 1
15 Leaf sheath disintegrating into fibres	absent 0, present 1

TABLE 2. Data matrix for systematic analysis of morphological and anatomical data of species of *Phoenix* with *Sabal mauritiiformis* as outgroup.

Species	1	2	3	4	5	6	7	8	9	10	11	12	13	14	15
Sabal mauritiiformis	0	–	0	0	–	–	–	0	0	0	0	0	0	0	0
Phoenix acaulis	1	0	1	1	0	0	1	0	0	1	1	0	1	0	1
P. andamanensis	0	0	0	0	1	0	0	0	1	0	0	0	0	0	1
P. caespitosa	1	0	1	1	0	0	0	0	0	1	1	0	1	0	1
P. canariensis	0	1	1	1	0	0	0	0	0	1	1	1	1	0	1
P. dactylifera	1	1	1	1	0	0	0	0	0	1	1	1	1	0	1
P. loureiri var. *loureiri*	1	0	1	1	0	0	0	0	0	1	1	0	1	1	1
P. loureiri var. *humilis*	1	0	1	1	0	0	0	0	0	1	1	0	1	0	1
P. paludosa	1	0	0	0	0	0	0	1	0	0	0	0	1	1	1
P. pusilla	1	0	1	1	0	0	0	0	0	1	1	0	1	0	1
P. reclinata	1	0	0	0	0	1	0	0	0	1	0	0	0	0	1
P. roebelenii	1	0	1	1	0	1	0	0	0	0	0	0	0	0	1
P. rupicola	0	0	0	0	1	0	0	0	0	0	0	0	0	0	1
P. sylvestris	0	1	1	1	0	0	0	0	0	1	1	1	1	0	1
P. theophrasti	1	1	1	1	0	0	0	0	0	1	1	1	1	0	1

Analysis

Morphological and anatomical data were analysed separately, and in combination with molecular data using the heuristic search option of PAUP version 3.1.1 (Swofford 1990) installed on a Power Mackintosh 8100/80. Maclade version 3.01 (Maddison & Maddison 1992) was used to optimise character states on trees. Both PAUP and Maclade version 3.01 were used to evaluate character fit on trees and to generate cladograms for publication.

Results

Cladograms resulting from three separate analyses of morphological and anatomical data in isolation, molecular data, and all three in combination are given in Figures 1 – 3.

DISCUSSION

Morphological versus molecular data

In comparison with separate systematic analyses of morphological and anatomical data (Fig. 1), and molecular data (Fig. 2), the combination of all three data sets (Fig. 3) results in a loss of resolution and an increase in the number of equally parsimonious trees generated, thus indicating conflict between the data sets in their support of a pattern of species relationships within *Phoenix*. Incongruence between the data sets centres upon the positions of *P. reclinata* and *P. pusilla*. Morphological and anatomical data support *P. reclinata* as associated with *P. paludosa*, *P. andamanensis*, *P. rupicola* and *P. roebelenii*, and *P. pusilla* with a different clade of species. Molecular data place *P. reclinata* and *P. pusilla* in unresolved positions.

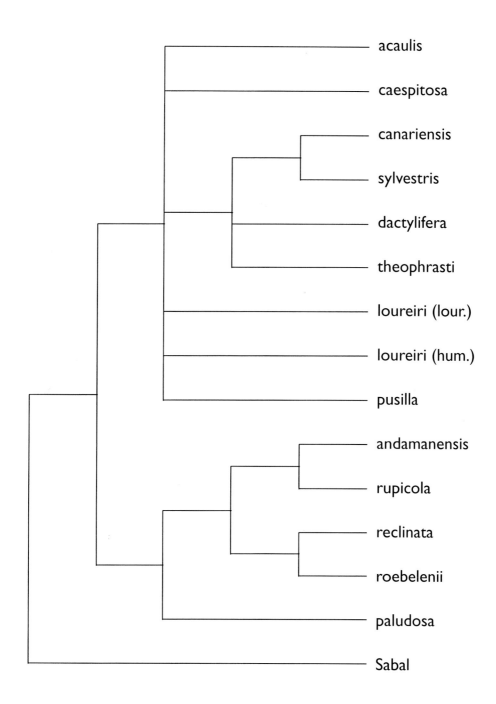

FIG. 1. One tree generated by systematic analysis of morphological and anatomical data for all species of *Phoenix* with *Sabal mauritiiformis* included as outgroup. Tree length = 17, CI = 0.647, RI = 0.812.

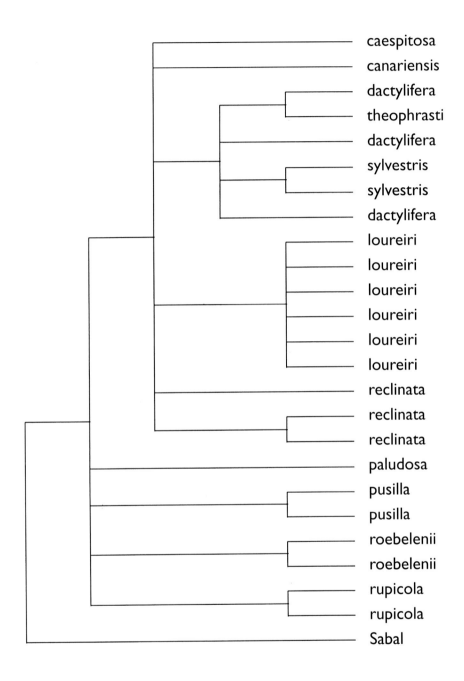

Fig. 2. Strict consensus of 72 equally parsimonious trees generated by systematic analysis of *5S* spacer sequence data from 24 samples of *Phoenix* representing all but two species of the genus. *Sabal mauritiiformis* is included as outgroup. Tree length = 100, CI = 0.700, RI = 0.815.

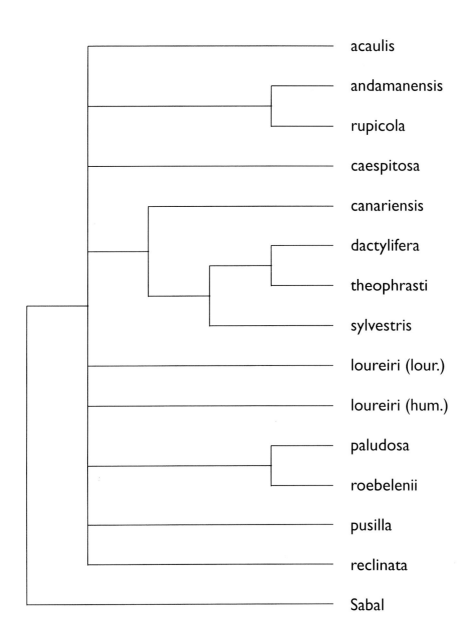

FIG. 3. Strict consensus of 54 equally parsimonious trees generated by combined analysis of morphological, anatomical and 5S spacer sequence data for all species of *Phoenix* with *Sabal mauritiiformis* included as outgroup. Tree length = 98, CI = 0.592, RI = 0.623.

Species relationships

Data incongruence resulting in poor resolution of the combined data consensus tree limits what can be inferred about species relationships within *Phoenix*. As a result I consider any biogeographical discussion of the genus to be premature. For a majority of species a discussion of relationships and affinities is speculative, based on support from either morphological or molecular data. However, both data types agree in their support of a clade comprising the large tree palm species; *P. dactylifera*, *P. theophrasti* and *P. sylvestris*, supporting conclusions proposed in the taxonomic account.

Support for *P. rupicola* and *P. andamanensis* as sisters is based on morphological data alone since a 5S spacer sequence is currently lacking for the latter species. However, on the basis of their very close morphological similarity I would expect 5S spacer sequence data to support them strongly as sisters. The biogeographical basis of the close relationship between these two species of limited distribution, *P. rupicola* from Bhutan and northeastern India, and *P. andamanensis* from the Andaman Islands, has not been thoroughly investigated and needs further study. Close affinity of *P. paludosa* with *P. roebelenii* is supported by both data types. Morphological and anatomical data support the grouping of *P. paludosa* and *P. roebelenii* in the same clade; however, their relationship as sisters, as indicated by combined analysis, is more a result of strong molecular support.

TAXONOMY

History of the genus

The earliest references to palms assignable to *Phoenix* concern the date palm, *P. dactylifera*, and its cultivation in ancient times by the Babylonians, Assyrians, Phoenicians and Egyptians, in Arabia and the Near East. The oldest known written records are found in Babylonia, dating to about 4000 BC, and these tell of date palm cultivation already well advanced. Popenoe (1924, 1973) gave a clear account of the date palm in antiquity. References are also given in classical literature. Theophrastus in *Enquiry into Plants* (370 – 285 BC, see Hort 1916) recorded multi-stemmed palms in the Mediterranean, currently attributable to *Phoenix theophrasti* (Greuter 1967). Pliny in *Natural History* (see Rackham 1945) discussed aspects of date palm cultivation in Arabia and the Near East, and referred also to the palms of the Canary Islands as *Palmeta caryotas ferentia*, and to multi-stemmed palms in Syria, Egypt and Crete.

A pre-Linnean student of the date palm was Engelbert Kaempfer. Gibbon (1776 – 1788) in *Decline and Fall of the Roman Empire* noted that 'The learned Kaempfer as a botanist, an antiquary and a traveller, has exhausted the whole subject of palm trees'. In an early series of illustrations, Kaempfer (1689) depicted aspects of date palm inflorescences and manual pollination techniques. The culmination of his work on date palms was published as *Phoenix persicus* or *A History of the Date Palm*, an extensive account of date palm botany, cultivation, history and culture (Kaempfer (1712), see Muntschick (1987) for translation). This account was known to Linnaeus who specifically cited cultivated *Palma hortensis* of Kaempfer (1712) in his accounts of *Phoenix* in both *Flora Zeylanica* (Linnaeus 1747) and *Species Plantarum*

(Linnaeus 1753). The illustrations of Kaempfer (1712) were identified as lectotypes for the name *Phoenix* by Moore & Dransfield (1979).

The pre-Linnean names *Hindindi* and *Mahaindi* of Hermann (1698, 1717) and *Katou-indel* of Rheede (1678 – 1703) are all referable to *Phoenix* and have been the cause of much taxonomic confusion. Hermann (1698, 1717) described *Hinindi* and *Mahaindi* as palms of Sri Lanka and southern India. *Katou-indel* was described by Rheede (1678 – 1703) in *Hortus Indicus Malabaricus* as a palm from India. The description of *Elate sylvestris* in *Musa Cliffortiana* (Linnaeus 1736) was based entirely upon *Katou-indel*. In later accounts, Linnaeus grouped *Katou-indel* with *Hinindi* of Hermann, as *Vaga* in *Flora Zeylanica* (Linneaus 1747) and as *Elate* in *Species Plantarum* (Linneaus 1753). Roxburgh (1832) transferred *Elate* to *Phoenix* but made no reference to two components in *Elate*. Hamilton (1827) acknowledged the separate elements but was not clear in his diagnosis on how they should be treated taxonomically. Martius (1823 – 53) finally clarified the nomenclatural confusion by limiting *Elate sylvestris* (and thus *Phoenix sylvestris*) to *Katou-indel* and transferring *Hinindi* to P. *pusilla* Gaertn.

Post-Linnean generic names attributable to *Phoenix* are *Palma* Mill., *Dachel* Adans. and *Phoniphora* Neck. (Moore 1963b). Miller (1754) validly published *Palma* in describing a mixed group of 21 species, referable to current species of *Draceana* Vand., *Zamia* L. and eight palm genera. *Palma dactylifera* (L.) Mill. is referable to *Phoenix dactylifera*. Miller's failure to adopt a published name previously proposed by Linnaeus (1753) makes the name *Palma* illegitimate (Moore 1963b). Similarly, *Dachel* Adans. and *Phoniphora* Neck. are illegitimate names, treated as valid synonyms of *Phoenix*.

The last monograph of *Phoenix* was written by Odoardo Beccari (1890). In his critical treatment Beccari included descriptions of ten species of *Phoenix*, and a preliminary key to identification. Other studies of *Phoenix* have been entirely floristic (e.g., Griffith 1845; Roxburgh 1832; Beccari & Hooker 1892 – 93; Gamble 1902; Blatter 1926; Mahabalé & Parthasarathy 1963), rather than monographic. Since the monograph of Beccari (1890) three species have been described as new: *P. caespitosa* Chiov. from Somalia, *P. atlantica* A. Chev. from the Atlantic Islands and *P. theophrasti* Greuter from Crete. Moore (1963a) in *An Annotated Checklist of Cultivated Palms* recognised a total of 12 species of *Phoenix*. The present treatment reduces two of these species to synonymy, includes a new species from the Andaman Islands and two varieties of *P. loureiri*, and considers a further species as incompletely known. Thirteen species of *Phoenix* are now recognised, with 33 synonyms and 57 nomina nuda, invalid or unpublished names treated as such.

Fossil record

As with all fossils, a certain amount of (palaeobotanic) confusion surrounds the naming of material and the attribution of it to extant taxa. Read & Hickey (1972) described key vegetative characteristics that can be used to assign fossils of sterile material to the palm family, but generic placement is more difficult. *Phoenix* is an exception. The unique induplicate pinnate leaves and lowest leaflets modified as acanthophylls allow fossil vegetative material to be unambiguously assigned to the genus. Using these characters Read & Hickey (1972) attributed fossil leaf material from the Eocene of France, previously described as *Palaeophoenix* Saporta, to *Phoenix*. Fossil material attributable to *Phoenix* dates to the Eocene (Reid & Chandler 1933; Chandler 1961 – 64; Machin 1971) and becomes increasingly common in the fossil record from this period onwards and is abundant in the European Oligocene, and the early Miocene of France, Switzerland and Croatia. Drude (1887) described fossil leaf material from Pleistocene deposits of the island of Santorini in the Aegean Sea as *Phoenix dactylifera fossilis* (details of this material are required before its identification is clarified). Pollen assigned to *Phoenix* is recorded from the London Clay and other British Eocene rocks (Reid & Chandler 1933; Chandler 1961 – 64; Machin 1971). Conwentz (1886) assigned a staminate palm flower preserved in the Baltic amber to *Phoenix*, but Daghlian (1978) expressed doubts on this identification on the basis of stamen morphology.

Seeds of *Phoenix* are also easily assignable, characterised as they are by the longitudinal raphe extending the full length of the seed. In Central Europe seed from the Lower Miocene was described as *Phoenix bohemica* by Buzek (1977), and from the Upper Eocene described as *P. hercynica* by Mai & Walther (1978). Of particular interest is fossil palm seed (*Phoenix*-like in character) from eastern Texas dating to the late Eocene to early Oligocene, discovered and described as *Phoenicites occidentalis* by Berry (1914). *Phoenicites* Brongn. now applies only to reduplicately-pinnate leaves (Read & Hickey 1972) and therefore would not now be used to describe fossil material attributable to *Phoenix*. A second report of a *Phoenix*-like taxon from the United States was given by Schmidt (1994) who considered petrified palm wood from the late Oligocene to early Miocene of Louisiana to be more like the wood of *Phoenix* than any other genus on the basis of stem structure and vascular bundle characteristics; however, the material cannot be unambiguously assigned to the genus.

Species concept

I used a morphological approach to recognising taxa in this monograph of *Phoenix*. I looked at the constancy of morphological character states both within and between populations and recognised as species only those smallest units which can be diagnosed by constant character states. Constancy within species was tested with molecular data. Systematic analysis of 5S spacer sequences, obtained from one population each of four species and several populations of seven species, indicated constancy (of 5S spacer sequences) within those species previously delimited by morphological data. Further molecular data may also provide support for discrete lineages within widespread species such as *P. loureiri* and *P. reclinata*. However, in my opinion intraspecific taxa should have morphological support and lineages supported by molecular data alone should not be given formal taxonomic recognition. The polymorphic *P. loureiri* and *P. reclinata* have been the cause of taxonomic problems because of description of poorly-defined infraspecific taxa. I found consistent morphological support for recognition of two varieties within *P. loureiri* but recognised none within *P. reclinata*. A bias of fieldwork towards Asia has given my knowledge of morphological variation within species of *Phoenix* a somewhat Asian bias at the expense of variation in African taxa; fieldwork in Africa may provide morphological data (not evident in the herbarium) which support recognition of infraspecific taxa within *P. reclinata*.

TAXONOMIC ACCOUNT

Phoenix *L.*, Sp. Pl.: 1188 (1753); Cav., Icon. 2: 12 – 15 (1793); Willd., Linn. Sp. Pl. (ed. 4), 4(2): 730 – 731 (1806); Buch.-Ham., Trans. Linn. Soc. London 15: 82 – 89 (1827); Roxb., Fl. Ind. ed. 2: 783 – 790 (1832); Kunth, Enum Pl. 3: 254 – 258 (1841); Griff., Calcutta J. Nat. Hist. 5: 344 – 355 (1845); Mart., Hist. Nat. Palm. 3: 257 – 276 (1849); Griff., Palms Brit. E. Ind.: 136 – 147, pl. 228, A - 229, A, B (1850); Brandis, Forest Fl. N.W. India: 552 – 556 (1874); Becc., Malesia 3: 345 – 416, pl. 43, f. 1 – 3, 44, f. 1 – 6 (1890); Becc. & Hook. f., Fl. Brit. India 6: 424 – 428 (1892); Gamble, Man. Ind. Timb. (ed. 2): 730 – 732 (1902); Brandis, Indian Trees: 644 – 646 (1906); Becc., Webbia 3: 238 – 240 (1910); Becc., Bull. Mus. Hist. Nat. (Paris) 17: 148 – 160 (1911); L. H. Bailey, Stand. Cycl. Hort.: 2593 – 2594 (1916); Blatt., Palms Brit. Ind.: 1 – 43, pl. 2 – 9, f. 1 – 5 (1926); Magalon, Contr. Étud. Palmiers Indoch.: 20 – 30 (1930); Gagnep. & Conrard in Lecomte, Fl. Indo-Chine 6: 974 – 978 (1937); Vasc. & Franco, Portugaliae Acta Biol., Sér. B, Sist. 2: 307 – 318 (1948); Mahab. & Parthasarathy, J. Bombay Nat. Hist. Soc. 60 (2): 371 – 387 (1963); N. W. Uhl & J. Dransf., Genera Palmarum: 214 – 217 (1987). Type species: *P. dactylifera* L. Lectotype: *Palma hortensis mas et foemina* Kaempf., Amoen. Exot. Fasc. 668, 686, t. 1, 2 (1712) (see Moore & Dransfield 1979).

Vaga L., Fl. Zeyl.: 187 (1747). No type designated.

Elate L., Sp. Pl.: 1189 (1753). Type species: *E. sylvestris* L. = *Phoenix* sylvestris (L.) Roxb. Lectotype: *Katou-indel* Rheede, Hort. Malab. 3: 15 – 16, pl. 22 – 25 (1678 – 1703); Mart., Hist. Nat. Palm. 3: 270, 273 (1845). The names *Phoenix* and *Elate* were published in *Species Plantarum* (Linnaeus 1753); the former has been chosen to take precedence over the latter (see Moore & Dransfield 1979) due to its greater familiarity.

Palma (L.) Mill., Gard. Dict. Abr. ed. 4, *nom. illegit.* (1754). Lectotype: *P. dactylifera* (L.) Mill., Gard. Dict. ed. 8 (1768). See Moore (1963b).

Dachel Adans., Fam. Pl. 2: 25, 548 (1763). No type designated.

Phoniphora Neck., Elem. Bot. 3: 302 (1790). No type designated.

Zelonops Raf., Fl. Tellur. 2: 102 (1837). Type: *Z. pusilla* (Gaertn.) Raf. (*Phoenix pusilla* Gaertn.). *Zelonops* was described as a small palm of India and Vietnam (probably *P. loureiri* Kunth) by Rafinesque (1837) who misunderstood the floral characteristics of *P. dactylifera*, and did not consider it to belong in *Phoenix*.

Palaeophoenix Saporta, Ann. Soc. Agric. Puy 33: 25 (1878). Type: *P. aymardi* Saporta, ibid., pl.1. Eocene: near Puy-en-Velay, France.

Dwarf to large, solitary or clustered, armed, pleonanthic, dioecious palms. *Stems* to 30 m tall or less than 10 cm high and bulbiferous, often with persistent leaves below crown; internodes short and congested with persistent, spirally-arranged, diamond-shaped leaf-bases, or internodes elongate, c. 5 – 10 cm apart and leaf bases not persistent. *Leaves* induplicately pinnate, persistent below crown, otherwise marcescent; sheath reddish-brown to brown-black, fibrous; true petiole absent to very short; rachis elongate, tapering, adaxially rounded or flat to angled, abaxially rounded to flat, usually terminating in a leaflet; proximal rachis with leaflets modified as acanthophylls. *Leaflets* regularly arranged or variously fascicled, in one to four planes of orientation, single-fold, acute to acuminate, flaccid or stiff, sometimes decurrent along rachis; lamina with no true midrib but midvein often prominent abaxially, adaxial surface glabrous, abaxial surface glabrous or with discontinuous scurfy, white ramenta or with veins darkened with tannin-filled cells. *Staminate inflorescences* interfoliar, erect, compact, branching to one order, not extending far beyond prophyll; prophyll coriaceous or papery, splitting one to two times between margins; peduncle flattened; rachillae crowded along rachis, flexuose. *Staminate flowers* along full length of rachillae, yellow-white, sweet to musty smelling; calyx a 3-lobed cupule, lobes variously distinct; petals 3 (rarely 4), valvate, much exceeding calyx in length, with margins variously jagged and apices rounded or acute to acuminate; stamens 6 (rarely 7 – 9) with minute filaments and anthers linear-latrorse, yellow-white to yellow-brown; pistillodes 3, minute; pollen elliptic or circular, monosulcate, with reticulate or finely reticulate, tectate exine. *Pistillate inflorescences* interfoliar, erect, arching or becoming pendulous at fruit maturity; prophyll bivalved, 2-keeled, coriaceous or papery, glabrous or floccose with reddish-brown tomentum; other bracts inconspicuous; peduncle flattened, yellow-green to orange, variously elongated; rachillae unbranched, numerous, irregularly arranged or in loose spirals along rachis. *Pistillate flowers* mostly in distal half to two thirds of rachilla, globose, yellow-white to pale green, each subtended by a non-persistent papery bract; calyx a 3-lobed cupule, lobes variously distinct; petals 3, rarely 4, imbricate, about twice as long as calyx; staminodes 6, scale-like; carpels 3, rarely 4, usually only one developing to maturity, follicular, ovoid, narrowed into a short, recurved, exserted stigma; ovule attached adaxially at the base, anatropous. *Fruit* ovoid to oblong or elongate, with stigmatic remains apical; epicarp smooth and mesocarp fleshy, endocarp membranous. *Seed* elongate, terete or plano-convex in shape, deeply grooved with intruded testa below the elongate raphe, hilum basal, rounded; embryo lateral opposite raphe or basal; endosperm homogeneous or ruminate. Germination remote-tubular; eophyll undivided, narrowly lanceolate; n = 18 (16).

KEY TO THE SPECIES

1. Leaflets **either** with surfaces discolorous and abaxial surface pale with tannin-stained veins, **or** surfaces concolorous and abaxial surface with white ramenta along midrib and/or veins · 2
   Leaflets with surfaces concolorous and abaxial surface without abaxial ramenta or tannin-stained veins · 6
2. Stems to 2 (3) m tall with diamond-shaped persistent leaf bases throughout, each with a central 'bump' of remnant vascular tissue, internodes very short. Southern China and northern regions of Indochina · · · · · · 1. **P. roebelenii**
   Stems to 10 (12) m tall, with leaf bases present only below crown, partially or completely ringed by narrow leaf base scars, internodes elongate · · · · · · · 3
3. Stems up to 8 cm in diameter, reddish-brown, completely ringed by regular leaf base scars. Leaflets discolorous, more or less irregularly arranged in more than one plane. Abaxial lamina surface greyish, veins darkened with tannin, without persistent white ramenta. Fruit flattened dorsiventrally, embryo basal. Coastal regions from Bay of Bengal eastwards to Malay Peninsula and northern Sumatra · 2. **P. paludosa**

Stems to 20 cm in diameter, pale to greyish-brown, partially or completely ringed by oblique, narrow leaf base scars. Leaflets concolorous, closely and regularly arranged in one plane. Abaxial lamina surface green, veins not darkened with tannin, with persistent white ramenta in the midrib region and/or along veins. Fruit circular in transverse section, embryo lateral · · · 4

4. Clustering, stems to 10 (12) m. Staminate petal apices acute to acuminate and with jagged margins. Africa and Madagascar · · · · · · · · · · · · 3. **P. reclinata**

    Solitary, stems to 3 – 5 m. Staminate petals obtuse with smooth margins · · · · 5

5. Endosperm homogenous. Northeastern India, and southern and central districts of Bhutan · 4. **P. rupicola**

    Endosperm ruminate. Andaman Islands · · · · · · · · · · · · · 5. **P. andamanensis**

6. Robust tree palms up to 30 m tall or stemless shrubs. Leaflets very stiff, glaucous, pale when dry. Mediterranean, North Africa, Arabian Peninsula, Near East · 7

    Moderate tree palms up to 5 m tall or stemless shrubs. Leaflets flaccid, leathery or moderately stiff. Asia · 11

7. Solitary palms. Pseudopetiole with acanthophylls conspicuously conduplicate (folded), congested at leaf base. Fruits ripening golden-brown. India and Canary Is · 8

    Clustering palms, often with basal suckers. Pseudopetiole with acanthophylls spine-like, sparsely arranged. Fruits variable, but mostly ripening deep red to purplish-brown to black. North Africa, Arabian Peninsula and the Near East · 9

8. Trunk to 15 m tall and to 75 (100) cm in diameter. Leaflets to 200 on each side of rachis, closely and regularly inserted in one plane of orientation. Canary Islands · 6. **P. canariensis**

    Trunk to 15 (20) m tall and to 20 – 30 cm diameter. Leaflets to 100 on each side of rachis, irregularly arranged in more than one plane of orientation. India, Pakistan. · 7. **P. sylvestris**

9. Clustering palms forming extensive thickets. Stems less than 1 m tall. Arabian Peninsula and Horn of Africa · · · · · · · · · · · · · · · · · · 8. **P. caespitosa**

    Clustering palms or solitary stem with basal suckers. Stems to 30 m tall. Mediterranean, North Africa, Arabian Peninsula, the Near East · · · · · · · 10

10. Fruits to 4 cm long with thick, fleshy mesocarp. Seeds with pointed apices. Cultivated widely in North Africa, Arabian Peninsula and the Near East · 9. **P. dactylifera** (cultivated)

    Fruits less than 2 cm long with sparse, mealy mesocarp. Seeds with rounded apices. Crete, Turkey, Near East · · · · · 10. **P. theophrasti** and 9. **P. dactylifera** (feral)

11. Acaulous. Leaflets to 1.5 cm wide with conspicuous marginal veins. Infructescences borne at ground level, hidden amongst leaves, peduncle to 15 cm long. Mature fruit congested in arrangement along rachillae, each one subtended by a prominent swelling of the rachilla. Apical stigmatic remains prominently pointed. Northern India and Nepal · · · · · · · · · · 11. **P. acaulis**

    Stems eventually to 4 (5) m tall, after a long initial stemless period. Leaflets to 2.5 cm wide with marginal veins not distinct. Pistillate inflorescence elongating on fruit set, peduncle up to 1.5 m long. Mature fruit restricted to upper half to two thirds of rachilla length, each not subtended by a bractiform swelling of the rachilla. Apical stigmatic remains not prominently pointed. India, Sri Lanka eastwards to Far East · 12

12. Leaflets four-ranked, often drying very pale green, sub-spathulate in shape with very sharp, needle-like apices. Staminate petals ovate, up to 4 × 4 mm, with apex entire. Seed glossy, chestnut-brown, up to 10 × 5 mm. Intrusion of testa into endosperm in the region of the raphe often Y-shaped in transverse section. Sri Lanka · 12. **P. pusilla**
Leaflets arranged in more than one plane but not four-ranked, not drying pale green, elongate, apices not sharp. Staminate petals oblong-elongate, up to 6 × 2.5 mm, with apex uneven, roughly undulate and often thickened. Seed matt grey-brown, up to 18 × 10 mm. Intrusion of testa into endosperm in the region of the raphe not elaborated, circular in transverse section. India eastwards to the Far East · 13

13. Leaflets with continuous strip of tannin-filled cells along margin, and discontinuous patches of such cells abaxially in the midrib region (visible with a hand-lens). Indo-China, southern China, Hong Kong, Taiwan and Batan and Sabtang Islands of the Philippines · · · 13a. **P. loureiri** var. **loureiri**
Leaflet margin and abaxial midrib region without tannin-filled cells. India · 13b. **P. loureiri** var. **humilis**

1. Phoenix roebelenii *O'Brien*, Gard. Chron., ser. 3, 6: 475, f. 68 (1889); C. Roebelen, Gard Chron., ser. 3: 758 (1889); Becc., Webbia 3: 237 – 245 (1910) & Bull. Mus. Nat. Hist. (Paris) 17: 148 – 160 (1911); L. H. Bailey, Stand. Cycl. Hort.: 2594, f. 2918, 2919 (1916); A. Chev., Rev. Int. Bot. Appl. Agric. Trop. 3: 837 – 839 (1923); Magalon, Contr. Étud. Palmiers Indoch.: 24, pl. 1 – 2, f. 1 (1930); Gagnep. & Conrard in Lecomte, Fl. Indo-Chine 6: 946 – 1056 (1937); Vasc. & Franco, Portugaliae Acta Biol., Sér. B, Sist. 2: 317, figs. 5, 19-5 (1948); H. E. Moore, Baileya 1 (2): 25 – 30, f. 14 – 15 (1953); H. E. Moore, Principes 7 (4): 157 (1963); S. J. Pei & S. Y. Chen, Fl. Reipubl. Pop. Sin. 13(1): 6 – 11 (1991); I. Hoffman, Palm J. (March): 17 – 19 (1994); S. Barrow, Principes 38 (4): 177 – 181 (1994). Holotype: Laos, R. Mekong, Oct. 1889 (ster.), *O'Brien* s.n. (K!).

Clustering palms (often solitary in cultivation), forming clumps with stemless plants suckering at base of taller stems. *Stem* 1 – 2 m (rarely 3 m) high, without sheaths to 10 cm in diam., erect or twisted, pale, becoming smooth with age, marked with diamond-shaped persistent leaf bases each with a central bump of remnant vascular tissue; stem base developed with a root boss; roots occasionally emerging from stem above ground level. *Leaves* arching, 1 – 1.5 (2) m long; pseudopetiole to c. 50 cm long; leaf sheath reddish-brown, fibrous; acanthophylls arranged singly or paired, c. 12 on each side of rachis, orange-green, to 8 cm long; leaflets regularly arranged, opposite, c. 25 – 50 on each side of rachis, linear, concolorous, deep green, often flaccid, to 40 × 1.2 cm; lamina with discontinuous white scurfy ramenta along abaxial veins and midrib, almost totally covering abaxial surface of unexpanded (sword) leaves. *Staminate inflorescences* pendulous; prophyll coriaceous, two-keeled, splitting once abaxially between keels, c. 30 – 60 cm long; peduncle to 30 cm long; rachillae 7 – 20 cm long. *Staminate flowers* with calyx a three-pointed cupule, 1.2 mm high, yellow-white; petals pale yellow-white with acuminate apices and with jagged margins, 7 – 8 × 2 – 2.5 mm; anthers 3.5 – 4 mm long. *Pistillate inflorescences* erect, arching as fruits ripen, up to 35 cm long; prophyll coriaceous, two-keeled, to 35 cm long × c. 5 cm wide, splitting once adaxially between keels to reveal inflorescence; peduncle green, to c. 30 × 3 cm; rachillae with bulbous bases, orange-green, occasionally branched to one order, subtended by papery bracts (c. 4 cm long). *Pistillate flowers* pale green, arranged in distal three quarters of rachilla, subtended by papery bracts to 5 mm long; calyx a three-pointed cupule, thickened and ridged up to apices, striate, 2 – 2.5 mm high; petals 3.5 × 4 mm with acute apices; generally only one carpel reaching maturity. *Fruits* obovoid, with persistent perianth, maturing from dark green to purplish brown, 13 – 18 × 6 – 7 mm; stigmatic remains apical, 1 mm long, orange-brown, often recurved. *Seed* narrowly elongate, terete, with rounded apices, 7 – 3 mm; embryo lateral opposite raphe; endosperm homogeneous.

DISTRIBUTION. Northern Laos (Nam Ou valley), Vietnam (Upper Black R. region near Lai-Chau), and southern China (Xisuangbanna region of Yunnan), most notably along the banks of the R. Mekong.

HABITAT AND ECOLOGY. Closely associated with riverside or cliff habitats where it grows as a rheophyte. The rheophytic habit is rare within the palm family (Dransfield 1992). The clustering habit of *P. roebelenii* may help it to survive flooding.

SELECTED SPECIMENS EXAMINED. CHINA. Yunnan, banks of Mekong R., Chiu-lung Chiang, 23 Feb. 1922 (photo.), *Rock* 2531 (E!). LAOS. R. Mekong, Oct. 1889 (ster.), *O'Brien* s.n. (K!, holotype); (ster.), *Magalon* 28 (P!). VIETNAM. no precise locality, (pist.), *Balansa* 4471 (FI-B!, P!), (ster.) *Balansa* 4877 (FI-B!, K!, P!).

VERNACULAR NAMES. THAILAND. Paam sipsong pannaa (Xishuangbanna palm), [Smitinand (1948)]. VIÊTNAM. Cha rang (Moyenne Region), [Magalon (1930)].

USES. Since its introduction to Europe, *P. roebelenii* has become a popular and widely cultivated ornamental palm and is now found in private and botanical gardens around the world.

CONSERVATION STATUS. A naturally restricted distribution, habitat loss and a horticultural trade in wild-collected plants may mean that the wild populations of *P. roebelenii* merit 'vulnerable' status, but further studies are needed. Demand for *P. roebelenii* as an ornamental is mostly met through seeds and offshoots from cultivated plants. However, Barrow (1994) suggested that collection of mature palms from the wild poses an increasing threat.

2. Phoenix paludosa *Roxb.*, Hort. Bengal.: 73 (non vidi) (1814), Fl. Ind. 3: 789, pl. 1193 (1832); Royle, Ill. Bot. Himal. Mts.: 397, nomen (1840); Kunth, Enum. Pl. 3: 256 (1841); Mart., Hist. Nat. Palm. 3: 272, t. 136 (1849); Griff., Calcutta J. Nat. Hist. 5: 353 (1845) and Palms Brit. E. Ind.: 144, t. 229B (1850); Hook. f., Himal. J. 2: 355 (1854); Kurz, Rep. Veg. Andaman Isl.: 50 (1870), J. Asiat. Soc. Bengal, Pt. 2, Nat. Hist. 43 (2): 202 (1874) & Forest Fl. Burma 2: 536 (1877); Brandis, Forest Fl. N.W. India: 556 (1874); Gamble, Man. Ind. Timb.: 419 (1881); Becc., Malesia 3: 410, fig. 6, f. 58 – 61 (1890); Becc. & Hook. f., Fl. Brit. India 6: 427 (1892); Brandis, Indian

Trees: 646 (1906); Becc., Webbia 3: 239 (1910); C. E. Parkinson, Forest Fl. Andaman Isl.: 263 (1923); Blatt., Palms Brit. Ind.: 21, pl. 7, f. 3 (1926); Magalon, Contr. Étud. Palmiers Indoch.: 20 (1930); H. E. Moore, Principes 7(4): 157 (1963); Mahab. & Parthasarathy, J. Bombay Nat. Hist. Soc. 60(2): 371 – 387 (1963); Whitmore, Palms of Malaya: 86 – 87 (1973); Kiew, Malayan Nat. J. 42 (1): 16 (1988); S. M. Mathew & S. Abraham, Principes 38 (2): 100 – 104 (1994). Lectotype: Roxburgh (1832), Fl. Ind. 3: 789, pl. 1193 (K).

P. siamensis Miq., Palm. Archip. Ind.: 14 (1868). No type designated.

Clustering palm, growing in dense clumps. *Stem* 2 – 5 m tall, without leaf sheaths to 5 – 8 cm in diam., reddish-brown, dead leaves persistent below crown to half way down trunk, otherwise leaves falling to reveal annular leaf-base scars; dead leaves falling to expose internodes (c. 5 cm long); stem base with exposed roots. *Leaves* 2 – 3 m long with pseudopetiole 70 – 98 cm long × 1.5 – 2 cm wide at junction with sheath; leaf sheath reddish-brown, becoming greyish-black, persistent as a stiff, fibrous network; acanthophylls arranged individually or paired, c. 11 – 19 on each side of rachis, green-brown, to 8 cm long, flattened, stiff, sharp, with bulbous bases; sharp transition between leaflets and acanthophylls; rachis abaxially rounded, green, speckled with green-brown indumentum, adaxially flattened; leaflets opposite distally, otherwise grouped in 3s or 4s in more than one plane of orientation, c. 34 on each side of rachis, 12 – 40 × 0.4 – 2.2 cm, flaccid, margins often recurved, discolorous, adaxially green, abaxially smokey-grey, with non-persistent ramenta, veins darkened with tannin, and midrib prominently keeled. *Staminate inflorescence* erect, not exceeding prophyll in length, compact like a stiff brush; prophyll splitting twice between margins, c. 20 – 40 × 3 – 5 cm; peduncle c. 20 – 30 cm long; rachis crowded with 35 – 50 rachillae; rachillae 5 – 10 cm long. *Staminate flower* with calyx a three-pointed cupule 1 mm high; petals yellow, 6 – 8 × 2 – 2.5 mm; anthers yellow-brown. *Pistillate inflorescence* erect, opening out and elongating on fruit set; prophyll splitting twice between margins, c. 30 – 40 cm × 3 – 5 cm; peduncle to 30 cm × 0.7 – 1.5 cm; rachillae c. 18 × 37 in number, 9 – 30 cm. *Pistillate flower* with calyx cupule up to 2 – 2.5 mm high; petals 2 – 3 × 2 mm. *Fruit* ovoid-ellipsoid, with persistent perianth, maturing from yellow-green to orange to blue-black, 10 – 12 × 7 – 10 mm, with stigmatic remains sub-apical to 0.5 mm long. *Seed* ovoid with rounded apices, 9 – 11 × 7 – 5 mm, with raphe not extending full length of seed; embryo basal; endosperm homogeneous.

DISTRIBUTION. Coastal regions from the Bay of Bengal, Andaman and Nicobar Islands, Indochina, to Malay Peninsula and northern Sumatra.

HABITAT AND ECOLOGY. *Phoenix paludosa*, so distinctive within the genus in its morphology and anatomy, is also unique in its habitat. The species is found growing in pure stands at the edges of mangrove and in estuarine coastal swamps. It does not grow in true mangrove forest, but in surrounding areas periodically inundated by brackish water (Kiew 1988). In its native habitat, *P. paludosa* flowers from February to April and fruits ripen from June to December.

SELECTED SPECIMENS EXAMINED. BANGLADESH. Chittagong, 5 Nov. 1920 (pist.), *Cowan* 753 (E!). INDIA: ANDAMAN Is. Port Monat, 10 June 1893 (pist. naked), *King* s.n. (BM!, FI!); Alexandra Is., 6 March 1904, *Rogers* s.n. (K!). NICOBAR IS. Jan. 1893 (stam.), *King* s.n. (BM!, K!). ORISSA. Dharma, 24 April 1960 (pist.), *Srivastava* 69819 (DD!). WEST BENGAL. Sunderbuns, Jan. 1880 (stam.), *Gamble* 7733 (K!). PENINSULAR MALYASIA. Perak, Dindings, Lumut, 4 Aug. 1936, *Henderson* 31483 (K!). MYANMAR. 21 March 1906 (pist.), *Lace* 2965 (E!). SUMATRA. Batu Bahra, (pist.), *Yates* 1876 (BM!, P!). THAILAND. Khong La Un, north of Ranong, 25 April 1972 (pist.), *Chamlong & Whitmore* 3167 (BKF!, K!); Trang Province, Jaupa Distr., Kantang Forest Ecology Research Station, 22 Jan. 1994 (ster.), *Barrow & Tingnga* 28 (BKF!, K!). VIÊTNAM. Nha-Trang, 12°15'N, 109°10'E, March 1911 (stam.), *Robinson* 1247 (FI-B!, K!).

VERNACULAR NAMES. INDIA. Hintala (Sanskrit), hintal, golpatta (Bengal), hantal [Sunderbuns, *Chaffey* 100 (K!)], giruka tati (Telinga). MYANMAR. Thin-boung. CAMBODIA. Peng [Magalon (1930)]. PENINSULAR MALAYSIA. Dangsa (Penang), Blatter (1926). THAILAND. Peng-tha-le ('Sea *Phoenix*'), [Smitinand (1948)]; Mangrove Date Palm (English), [Whitmore (1973)]. VIÊTNAM. Cay cha la rieng, [Magalon (1930)].

USES. Leaflet fibre for rope and thatch (Blatter 1926); stems for walking sticks, rafters (Blatter 1926), and fence posts. Roxburgh (1832) reported that in the Sunderbuns there is a belief that snakes will avoid any person carrying a staff made from the stem of *P. paludosa*.

CONSERVATION STATUS. *Phoenix paludosa* is not considered to be a threatened species, although its conservation status varies regionally. Kiew (1988) conferred vulnerable conservation status upon the species in Peninsular Malaysia, reporting that all populations in the peninusla are at risk from increasing urban development, and several have been lost already. Certain populations in Thailand are under similar threats of drainage and urban expansion.

3. Phoenix reclinata *Jacq.*, Fragm. Bot. 1: 27, t. 24 (1801); Willd., Linn. Sp. Pl. (ed. 4), 4(2): 731 (1806); Spreng., Syst. Veg. 2: 138 (1825); Kunth, Enum. Pl. 3: 256 (1841); Mart., Hist. Nat. Palm. 3: 272, t. 164 (1849); Becc., Malesia 3: 349, t. 44, f. 1 (1890); Warb., Pflanzen. Ost-Afr.: 130 (1895); C. H. Wright in Oliv., Fl. Trop. Afr. 8: 103 (1901); Engl., Veg. Erde 2: 224, t. 10, fig. 149 (1908); L. H. Bailey, Stand. Cycl. Hort.: 2593 (1916); Blatt., Palms Brit. Ind.: 39, pl. 9, f. 5 (1926); Magalon, Contr. Étud. Palmiers Indoch.: 28 (1930); Jum. & H. Perrier, Fl. Madagascar 30: 19, fig. 3 (1945); R. O. Williams, Useful & Ornamental Plants in Zanzibar & Pemba: 411 (1949); Eggeling, Indig. Trees Ugan. Prot., ed. 1: 164, photo. 36 (1940); F. W. Andrews, Fl. Pl. Sudan 3: 304 (1956); Dale & Greenway, Kenya Trees & Shrubs: 12 (1961); H. E. Moore, Principes 7 (4): 157 (1963); T. A. Russell, Fl. W. Trop. Afr., ed. 2, 3: 169 (1968); Hamilton, Uganda For. Trees: 75 (1981); Troupin, Fl. Rwanda 4: 399, f. 170 (1987); J. Dransf., Fl. Trop. E. Afr., *Palmae*: 15, f. 1 (1986); M. F. Kinnaird, Conservation Biol. 6(1): 101 – 107 (1992); Beentje, Kenya Trees, Shrubs & Lianas: 644 (1994); Thulin, Fl. Somalia 4: 271 (1995); S. Barrow in J. Dransf. & Beentje, Palms of Madagascar: 47 (1995); P. Tuley, Palms of Africa: 16 (1995). Lectotype: t. 24 in Jacq., Fragm. Bot. (1801).

Phoenix spinosa Schumach. & Thonn., Beskr. Guin. Pl.: 437 (1827); Mart., Hist. Nat. Palm. 3: 275 (1849); A. Chev., Rev. Int. Bot. Appl. Agric. Trop. 32: 223 (1952).

Type: Ghana, *Thonning* 101 (C, holotype). *Phoenix spinosa* refers to a dwarf coastal form of the species. It is found growing in dense thickets along the Atlantic coast and is associated with higher rainfall regimes (Tuley 1995).

Phoenix abyssinica Drude, Bot. Jahrb. Syst. 21: 117 – 119 (1895); C. H. Wright in Oliv., Fl. Trop. Afr. 8: 102 (1901); Fiori, Boschire Piante Legnose dell Eritrea: 97, fig. 34 (1912); A. Chev., Rev. Int. Bot. Appl. Agric. Trop. 32: 217 (1952). Type: Eritrea, Tigré, Nov. 1861 (pist.), *Steudner* 1541 (FI-B! ex Herb. Berol.).

Phoenix reclinata var. *madagascariensis* Becc., Bot. Jahrb. Syst., Beibl. 87, 38: 4 (1906); Jum. & H. Perrier, Ann. Inst. Bot. Géol. Colon. Marseille 3, 1: 60 (1913); Becc., Palme Madagascar: 54, fig. 44 (1914). Lectotype: Loko-bé, Nossi-bé, Dec 1879 (stam.), *Hildebrandt* 3304 (BM!, K!, P!).

Phoenix comorensis Becc., Bot. Jahrb. Syst., Beibl. 87, 38: 5 (1906), Palme Madagascar: 54 (1914). Lectotype: Mayotte, 1850 (pist.), *Boivin* 3100 (K!, P!). *Phoenix reclinata* in Madagascar shows variation consistent with that of the species in Africa. It was first noted from Madagascar by Beccari (1890), who later (Beccari 1906, 1914) recorded it as *P. reclinata* var. *madagascariensis* Becc. on the basis of very minor differences in the persistent fruit perianth. Beccari (1906) separated *P. comorensis* from *P. reclinata* using similar characters. Jumelle & Perrier (1913, 1945) considered both taxa as synonymous with *P. reclinata*.

Phoenix dybowskii A. Chev., Fl. Afr. Centr., Énum. Pl. Récolt.: 331, *nomen* (1913); Rev. Int. Bot. Appl. Agric. Trop. 32: 224 (1952). Type: Central African Republic, Ungourras plateau, 14 Nov. 1902 (ster.), *Chevalier* 6152 (P!). Chevalier described *P. dybowskii* as a palm found hanging over riverbanks. I consider it merely to be a form of *P. reclinata*.

Phoenix reclinata var. *somalensis* Becc., Chiov., Res. Sci. Somalia Ital. 176, 230, Tab. 43, fig. 3,4 (1916). Lectotype: Uagadi near Bulo Nassib, 29 June 1913 (stam.), *Paoli* 430 (FT!).

Phoenix djalonensis A. Chev., Explor. Bot. Afrique. Occ. Franç.: 672, *nom.* (1920), Rev. Int. Bot. Appl. Agric. Trop. 32: 223, descr. (1952). Type: Guinea, Fouta Djallon, Kollangui, July 1906 (pist.), *Chevalier* 12823 (P!). Chevalier described *P. djalonensis* to refer to a tall, thin-stemmed palm from the uplands of the Fouta Djallon range. Tuley (1995) pointed out that this name could equally well be applied to all upland forms of *P. reclinata* throughout the continent.

Phoenix baoulensis A. Chev., Rev. Int. Bot. Appl. Agric. Trop. 32: 224 (1952). Type: Ivory Coast, Kodiokoffi Distr., Manikro, 6 Aug. 1909 (pist.), *Chevalier* 22314 (P!).

Clustering palm, often thicket-forming. *Stem* 10 (12) m, erect or oblique, without leaf sheaths to 20 cm in diam., dull brown, with persistent leaf sheaths 1 – 2 m below crown, otherwise becoming smooth, irregularly marked with oblique internode scars, cracked vertically; injured stem exuding clear yellowish gum. *Leaves* arcuate, c. 2 – 3.5 m long; leaf sheath fibrous, reddish-brown; pseudopetiole rounded abaxially, smooth, often channelled adaxially, to 50 cm long; acanthophylls irregularly arranged, often congested proximally, c. 10 – 15 on each side of rachis, 3 – 9 cm long; leaflets regularly arranged distally in one plane of orientation but median and proximal leaflets in fascicles of 3 – 5 and often fanned, c. 80 – 130 on each side of rachis, 28 – 45 × 2.2 – 3.6 cm; leaflet margin minutely crenulate; lamina concolorous, abaxial surface with white scurfy ramenta in midrib region. *Staminate inflorescence* erect; prophyll green-yellow in bud, strongly 2-keeled, coriaceous, splitting 1 or 2 times between margins, 40 – 60 × 5 – 6 cm; peduncle 10 – 30 × 1.3 cm, not greatly elongating beyond prophyll; rachis 17 – 30 cm; rachillae congestedly arranged in a narrow bush, numerous, 6 – 20 cm long. *Staminate flowers* creamy-white; calyx cupule 1 mm high; petals with apex acute-acuminate in shape and with jagged margins, 3 (rarely 4), 6 – 7 × 2 – 3 mm. *Pistillate inflorescence* erect, arching with weight of fruits; prophyll as for staminate inflorescence; peduncle green-yellow turning orange-brown, becoming pendulous on fruit maturity, to 60 – 1.5 cm; rachillae spirally arranged often in irregular horizontal whorls, c. 19 – 40 in number, to 6 – 55 cm long. *Pistillate flowers* usually only one carpel reaching maturity, 3 – 4 mm high. *Fruit* ovoid-ellipsoid or almost obovoid, ripening yellow to bright orange, 13 – 20 × 7 – 13 mm; mesocarp sweet, scarcely fleshy, c. 1 – 2 mm thick. *Seed* obovoid, with rounded apices, 12 – 14 × 5 – 6 mm; embryo lateral opposite raphe; endosperm homogeneous.

DISTRIBUTION. *Phoenix reclinata* occurs throughout tropical and subtropical Africa, northern and southwestern Madagascar and the Comoro Islands.

HABITAT AND ECOLOGY. *Phoenix reclinata* is a widely distributed species growing in a range of habitats, often seasonally water-logged or inundated, such as along watercourses, in high rainfall areas, in riverine forest, and even in rainforest areas (although always restricted to areas of sparse canopy). The species can also be found in drier conditions on rocky hillsides, cliffs and grasslands to 3000 m. The fruits of *P. reclinata* are animal-dispersed: their bright orange colour and sweet, slightly fleshy mesocarp is attractive to birds (parrots) (Schonland 1924), elephants (Corner 1966), lemurs (Petter *et al.* 1977), mangabey (forest monkeys) (Kinnaird 1992) and humans.

SELECTED SPECIMENS EXAMINED. ANGOLA. Lola Prov., Bibala, 14°46'S, 13°21'E, 1951 (stam.), *Teixeira* 497A (BM!). BENIN REPUBLIC. Goowe, 23 Feb. 1905, *Le Testu* 149 (BM!). BOTSWANA. Okavango R., 18°27'S, 22°30'E, 30 April 1975 (pist.), *Biegel et al.* 5042 (E!, K!). BURUNDI. Nkundarafe, Gitega-Ruyigi rd., 3°36'S, 30°06'E, 11 Oct. 1978 (pist., stam.), *Reekmans* 7146, 7147 (K!). CAMEROON. Mpalla, 12 km N of Kribi, 1968 (stam., pist.), *Bos* 3479 (K!). CENTRAL AFRICAN REPUBLIC. Ungourras Plateau, 14 Nov. 1952 (ster.), *Chevalier* 6152 (P!) [Type of *P. dybowskii*]. COMORO ISLANDS. Mayotte, (pist.), *Boivin* 3100 (K!, P!). ERITREA. Hamasen, 1570 m alt., 29 March 1909 (stam., pist.), *Fiori* 453 (FT!). ETHIOPIA. Kaffa Prov., 7 km E of Jimma, 1750 m alt., 7°40'N, 36°52'E, 26 Dec. 1961 (pist.), *Meyer* 7819 (K!). GABON. Maliba village, 24 Sept. 1969 (stam.), *Villiers* 365 (P!). GAMBIA. 1 March 1921 (stam., pist.), *Dawe* 70 (K!). GHANA. Aburi Hills, 5 Oct. 1899 (stam.), *Johnson* 459 (K!). GUINEA. Fouta Djallon, Mali region, Sept. 1954 (pist.), *Schnell* 7224 (P!). GUINEA-BISSAU. Mayombo, 11 May 1919 (stam., pist.), *Gossweiler* 8097 (BM!). IVORY COAST. Foro-Foro forest, c. 25 km N of Bouaké, 26 Sept. 1963 (stam., K!; ster., P!), *Oldeman* 402 (K!, P!). KENYA. Teita Distr., Ngerenyi, 1700 m alt., 17 Sept. 1953 (pist.), *Drummond & Hemsley* 4369 (FT!, K!). MADAGASCAR. 2 km S of Iharana, 13°28'S, 49°29'E, 24 June 1992, *Beentje & Andriampaniry* 4691, 4692 (K!). MALAWI. Blantyre Distr., Bangwe Hill, 4 km E of Limbe, 1260 m alt., 23 Nov. 1977 (pist.), *Brummitt et al.* 15156 (K!). MOZAMBIQUE.

Maputo, Inhaça Is., 13 Dec. 1984 (pist.), *Groenendijk & Dungo* 1587 (K!). NIGERIA. Muri Division, Mumye Distr., Gangoro Forest Reserve, 1350 m alt., 9 Feb. 1976 (pist.), *Chapman* 4127 (K!). RWANDA. Kibuye, 1978 (pist.), *Troupin* 15972 (K!). SAUDI ARABIA. Between Bani Sa'd and Jabal Ibrahim, Taif to Al Bahah road, 21 April 1988 (stam.), *Collenette* 6701 (K!). SENEGAL. Tambacounda region, Obadji to Kedougou road, 8 April 1993 (pist.), *Sambou et al.* 1569 (K!). SIERRA LEONE. No. 2 River Peninsula, 30 Sept. 1965 (stam.), *Morton & Jarr* 2298, 2395 (SL, K!, GC, WAG, FHI, IFAN). SOMALIA. Giuba R. near Bardera, 12 Nov. 1913 (pist.), *Paoli* 824 (FT!). SOUTH AFRICA. Natal Prov., Kosi Bay, 20 Jan. 1977 (pist.), *Balsinhas* 3099 (K!). TANZANIA. Tanga Distr., Lwengara R., 2.5 miles E of Korogwe, 300 m alt., 16 July 1953 (stam., pist.), *Drummond & Hemsley* 3342 (K!); N of Lembani on Korogwe-Moshi rd., 800 m alt., 15 Jan. 1976 (pist.), *Dransfield* 4832 (K!). UGANDA. Budongo Forest, 1200 m, alt., 28 Nov. 1938 (ster.), *Loveridge* 128 (K!). YEMEN. Al Jabin to Suq Ar Ribat, 700 m alt., 22 March 1984 (pist.), *Miller & Long* 5383 (E!). ZAÏRE. Katanga Prov., Pweto, July 1957 (pist.), *Devred* 3699 (K!). ZAMBIA. 8 km N of Mwinilunga, 1975 (pist.), *Brummit et al.* 14040 (K!). ZIMBABWE. Mutare, Odzani R. valley N of Penhalonga, 18°46'S, 32°41'E, 19 Dec. 1994 (pist.), *Wilkin* 724 (K!).

VERNACULAR NAMES. KENYA. mkindu (Swahili), gonyoorriya (Boni), meti (Digo), gedo (Ilwana), makindu (Kikuyu), sosiyot (Kipsigis), othith (Luo), ol-tukai (Maasai), konchor (Orma), itikindu (Sanya), alol (Somali), mhongana (Taveta), kigangatehi (Taita), nakadoki (Turkana), [Beentje (1994)]; mangatche [Kilimanjaro Distr., *Greenway* 3037 (K!)]. MADAGASCAR. Dara, taratra, taratsy, [Jumelle & Perrier (1913, 1945)]; calalou, [Morondava, *Grevé* 154 (P!)]. NIGERIA. Kajinjiri, dabino biri (Hausa), [Northern Prov., Zaria, *Conservator of Forests* s.n. (K!)]; deli (Fulani), kabba (Hausa), [Mambila Plateau, *Hepper* 1705 (K!)]. RWANDA. Umukindo, [Troupin (1987)]. SIERRA LEONE. Shaka-Le (Sherbro), kundi (Mende), [Bonthe Is., *Deighton* 2397 (K!)]. SOUTH AFRICA. Dikindu, makindu (Mbukushu), makerewa, shikerewa (Diriko), [Okavanga, *De Winter & Wiss* 4800 (K!)]. TANZANIA. Daro, taratra, mkindwi (Swahili), [Lamu Distr., *Dransfield* 4799 (K!)]; Luchingu (Fipa), [Mbugwe, *Bullock* 3074 (K!)]; kihangaga (Urukindu), [Lake Prov., *Tanner* 5845 (K!)]. UGANDA. Itchi (Madi), lukindu (Luganda, Lunyoro), musansa (Luganda, Busoga dialect), [Eggeling (1940)]; Wild Date Palm, enkinu (Luamba), emusogot (Ateso), ekingol (Karamojong), lukindu, mukindu (Luganda, Lunyoro, Lutoro), makendu (Lugisu), muyiti (Lugwe), otit (Luo. Acholi and Lango dialects), tit (Luo, Lango and Jonam dialects), itchi (Madi), kikindu (Lunyuli), lusansa (Lusoga), [Hamilton (1981)].

USES. All parts of *P. reclinata* palms are used for a range of purposes. Trunks are used as beams and poles in construction. Whole leaves are used as door entrances and covers, or fans for stoking fires and repelling insects. The leaf rachis is used for making thatch, floor mats and fish traps. It also forms a component of wattle for the construction of mud houses. Leaflets from sucker shoots are harvested for making baskets, hats, brushes, building ties, woven dolls and ornaments. The fruits are eaten as a snack and the seeds can be dried and ground into flour (Sierra Leone, *Deighton* 2397, K!). The palm heart is occasionally eaten as a vegetable. The sap is fermented into an alcoholic beverage and has been recorded as a remedy against urinary infections [Lake Prov., *Tanner* 5845 (K!)]. For a detailed study of the uses of *P. reclinata* in Tanzania see Kinnaird (1992).

CONSERVATION STATUS. Not threatened.

NOTES. The vegetative polymorphism of *P. reclinata*, which perhaps relates to ecological variation, has led to recognition of certain extreme phenotypes as distinct species or varieties (e.g., Chevalier 1952). This variation is such that delimitation of infraspecific taxa cannot be upheld by discrete characters.

4. Phoenix rupicola *T. Anderson*, J. Linn. Soc., Bot. 11: 13 (1869); Griff., J. Trav.: 46 (1847); Becc., Malesia 3: 395 (1890); Becc. & Hook. f., Fl. Brit. India 6: 425 (1892); Caruel, Gard. Chron.: 45, fig. 4 (June 1897); Gamble, Man. Ind. Timb.: 730 (1902); L. H. Bailey, Stand. Cycl. Hort.: 2593 (1916); Gamble, List of trees, shrubs & large climbers in Darjeeling Distr., Bengal: 86 (1922); Blatt., Palms Brit. Ind.: 14 (1926); H. E. Moore, Principes 7 (4): 157 (1963); Noltie, Fl. Bhutan 3 (1): 415 (1994). Type: India, West Bengal, Sivoka, Teesta valley, Feb. 1867, *Herb. Sikkimense T. Anderson* s.n. (CAL!, K!). Only one herbarium sheet in the Calcutta herbarium is definitely referable to T. Anderson. This sheet comprises sterile material only. Two fertile collections of *P. rupicola* held in the Calcutta herbarium, are possibly also those of T. Anderson.

P. rupicola var. *foliis argenteo varieg.* Hort. ex Rodigas, Ill. Hort.: 10, t. III (1887); Becc., Malesia 3: 397 (1890). Lectotype: t. III in Rodigas, Ill. Hort. 10 (1887).

Solitary tree palm. *Stem* 3 – 5 m high, without leaf sheaths up to 17 – 25 cm in diam., smooth with ill-defined internode scars. *Leaves* c. 1.5 – 2.5 m long; leaf sheath reddish-brown, fibrous; acanthophylls sparsely arranged in one plane of orientation, up to 10 – 15 on each side of rachis, often green and soft, to c. 7 cm long; leaflets closely and regularly inserted opposite in one plane of orientation, concolorous, dark glossy green, c. 24 – 60 × 1 – 3 cm; lamina abaxial surface with persistent, discontinuous white ramenta in midrib region. *Staminate inflorescence* erect; prophyll not seen; rachillae to c. 22 cm long. *Staminate flowers* not seen. *Pistillate inflorescence* erect, arching and becoming pendulous on maturity; prophyll not seen; peduncle 50 – 100 × 2.5 – 3.5 cm; rachillae arranged in horizontal fascicles, c. 120 in number, to c. 55 cm long. *Pistillate flowers* in upper half of rachilla length; calyx a 3-pointed cupule to 1.5 – 2.5 mm high; petals 3.5 × 4 – 6 mm. *Fruit* obovoid, 15 – 9 mm. *Seed* obovoid with squared apices, 12 – 15 × 5 – 7 mm; embryo lateral opposite raphe; endosperm homogeneous.

DISTRIBUTION. Southern (Samchi, Sankosh, Gaylephug and Deothang) and Central (Tongsa) Distrs. of Bhutan, and Darjeeling Distr. (Sivoka, Birick, Nimbong, Teesta and Mahanadi valleys) of West Bengal in India.

HABITAT AND ECOLOGY. Relatively inaccessible patches of warm, wet forest or more open areas on steep rocky hillsides, ravines and cliffs from 300 to 1220 m. Flowering in May and June; fruits ripe October – December.

SELECTED SPECIMENS EXAMINED. BHUTAN. Lower Mangde Valley, Tama, 27°10' N, 90°39'E, 26 June 1979 (stam.), *Grierson & Long* 1354 (E!); Samchi Distr., Khagra valley, near Gokti (26°49'N, 89°12'E), 2 March 1982 (ster.), *Grierson & Long* 3414 (K!). INDIA. ASSAM. Shillong, Kimin to Khunipahad, 25 Sept. 1959 (pist.),

Panigrahi 19485 (CAL!). SIKKIM. 24 June 1876 (ster.), *King* s.n. (BM!, CAL!); 19 Jan. 1877 (ster.), *Davis & Gamble* 2387a (CAL!); 21 Jan. 1877 (ster.), *Davis & Gamble* 2387b (K!). WEST BENGAL. Sivoka, Teesta valley, 23 Feb. 1867 (pist.), *Herb. Sikkimensis Anderson* s.n. (type CAL!, K!).

VERNACULAR NAMES. Takil (Nepalese); schap, sap, fam (Lepchas), [Gamble (1902)].

USES. Fruits of *P. rupicola* are sweet but mealy, and are eaten by mammals and birds. Gamble (1902) noted that the stem pith is eaten uncooked by local Lepcha people.

CONSERVATION STATUS. The conservation status of *P. rupicola* in its wild habitat is unclear. It has a limited range, making it vulnerable to habitat loss. The ability of *P. rupicola* to thrive in inaccessible habitats such as steep, rocky slopes, ridges and cliffs may help ensure its survival in the wild.

5. Phoenix andamanensis *S. Barrow* **sp. nov**. *P. rupicola*e affinis, sed endospermio ruminato non homogeneo differt. Type: Andaman Islands, North Andaman, Saddle Peak, 700 m alt., 14 Dec. 1990 (pist.), *Ellis* 14189 (K!).

Phoenix sp., Kurz, Rep. Veg. Andaman Isl.: 7, 50 (1870); Brandis, Indian Trees: 646 (1906); C. E. Parkinson, A Forest Flora of the Andaman Isl.: 263 (1923).

Solitary tree palm. *Stem* 1.5 – 3.5 (5) m, without leaf sheaths c. 15 cm diam. *Leaves* to c. 2.4 m long; acanthophylls sparsely arranged in one plane, to c. 4 cm long; leaflets closely and regularly inserted in one plane, 14 – 45 × 0.4 – 2.5 cm; lamina concolorous with discontinuous white, scurfy ramenta in midrib region on the abaxial surface. *Staminate inflorescence* with prophyll to c. 30 × 5 cm, coriaceous; rachillae to c. 10 cm long. *Staminate flowers* not seen. *Pistillate inflorescence* with prophyll splitting twice between margins, to 60 × 4 cm; peduncle to 100 × 1.2 cm; rachillae to c. 23 cm long. *Pistillate flowers* spirally arranged in distal half of rachilla, c. 20 in number; calyx cupule 1.5 mm high; petals 3 – 4 × 6 mm. *Fruit* oblong, to 19 × 10 mm, colour at maturity not known. *Seed* elongate, to 14 × 7 mm; embryo lateral opposite raphe, slightly supra-equatorial; endosperm ruminate.

DISTRIBUTION. *Phoenix andamanensis* has been recorded from one locality each in both North Andaman and Little Andaman, and from Cinque and Rutland Islands (Brandis 1906; Parkinson 1923). The modern distribution of the species is unknown.

HABITAT AND ECOLOGY. Higher ground (c. 450 – 700 m) on the islands. A recent report (Balachandran, *pers. comm.*) noted that the species occurs in undisturbed 'scrub jungle' on the eastern side of Rutland Island and northern end of North Cinque Island.

SELECTED SPECIMENS EXAMINED. ANDAMAN IS. NORTH ANDAMAN. Saddle Hill, 450 m alt., 28 Sept. 1905 (stam., pist.), *Osmaston* (CAL!); Saddle Hill, 500 m alt., 1 Dec. 1976 (pist.), *Balakrishnan & Nair* 4771 (CAL!); Saddle Peak, 700 m alt., 14 Dec. 1990 (pist.), *Ellis* 14189 (K!). RUTLAND IS. precise locality unknown, 13 Feb. 1904 (pist.), *Rogers* 132 (FI-B!, K!); Headland, North Dyer Point, 19 May 1904 (pist.), *Rogers* 285 (FI-B!, K!). CINQUE ISLAND. precise locality unknown, 7 April 1911 (stam., pist.), *Rogers* s.n. (CAL!); Cinque and Rutland Is., 20 July 1911 (seed), *Rogers* s.n. (K!). LITTLE ANDAMAN. Bumila Creek, Jan. 1903, *Rogers* s.n. (K!).

VERNACULAR NAMES AND USES. Not known.

CONSERVATION STATUS. The conservation status of *P. andamanensis* is unclear. It seems that the species was never common throughout the islands, but formed large populations in two localities. Brandis (1906) and Parkinson (1923) noted it as forming 'forests' on Cinque Island and north eastern Rutland Island. The fact that *P. andamanensis* was not found in a survey of palms of the Andaman and Nicobar Islands by Mathew & Abraham (1994) suggests that the species might now be rare.

NOTES. The existence of a second species of *Phoenix* in the Andaman Islands, in addition to *P. paludosa*, was noted by Kurz (1870), Brandis (1906) and Parkinson (1923), but its identity was not ascertained. Beccari provisionally named three herbarium specimens (*Rogers* s.n., 132 and 285 at K) of the species as *P. pusilla* var. *andamanensis* (nom. in sched.), but Brandis (1906) compared it with *P. rupicola*. I have found *P. andamanensis* to be similar morphologically and anatomically to *P. rupicola*. Both species are solitary in habit and have broad leaflets (to 3 cm in width) which are closely and regularly inserted in one plane of orientation. The abaxial lamina surface of both species bears discontinuous, abaxial white ramenta in the midrib region. Despite similarities between *P. andamanensis* and *P. rupicola*, the former is immediately distinguished by its seed with ruminate endosperm. The close relationship between *P. andamanensis* and *P. rupicola* supports the acknowledged similarity between the flora of the Andaman Islands with that of northeast India. Rao (1996) cited two rare orchid species from northeastern India, *Porpax meirax* King & Pantl. and *Ascocentrum ampullaceum* Schltr., which are also found on Saddle Peak on North Andaman.

6. Phoenix canariensis *Chabaud*, La Provence Agricole et Horticole Illustrée 19: 293 – 297, fig. 66 – 68 (1882); Naudin, Rev. Hort. 57: 541 (1885), Rev. Hort. 60: 180 (1888), Ill. Hort. 33: 8 (1886); Becc., Malesia 3: 369, fig. 17, t. 43, 2, f. 15 – 21 (1890); L. H. Bailey, Stand. Cycl. Hort.: 2594 (1916); Blatt., Palms Brit. Ind.: 41, f. 4 (1926); Vasc. & Franco, Portugliae Acta Biol., Sér. B, Sist. 2: 312, figs. 3, 19-2 (1948); A. Chev., Rev. Int. Bot. Appl. Agric. Trop.: 219 (1952); H. E. Moore, Principes 7 (4): 157 (1963), Principes 15 (1): 33 – 35 (1971) and Fl. Vitiensis Nova 1: 401 (1979); D. Lüpnitz & M. Kretschmar, Palmarum Hortus Francofurtensis 4: 23 – 63 (1994). Type: *P. canariensis* was described from cultivated plants without designation of a type. Lectotype: Figs. 66 – 68 of Chabaud, La Provence Agricole et Horticole Illustrée 19: 293 – 297 (1882), chosen by Moore (1971a).

P. dactylifera var. *jubae* Webb & Berthel., Hist. Nat. Iles Canaries 3 (2): 289 (1847). No type designated.

P. cycadifolia Hort. Athen. ex Regel, Gartenflora 28: 131, pl. 974 (1879); Moore, Principes 15 (1): 33 – 35 (1971), *nom. utique rej. prop.*

P. jubae (Webb & Berthel.) D. H. Christ, Bot. Jahrb. Syst. 6: 469, *nom.* (1885), 9: 170, *nom.* (1888). No type designated.

P. canariensis var. *porphyrococca* Vasc. & Franco, Portugaliae Acta Biol., Sér. B, Sist. 2: 313, fig. 19 – 23 (1948). Type: *Palma Horto Botanico Olissiponense*, Portugaliae Acta Biol., Sér. B, Sist. 2: 313, fig. 19 – 23 (1948).

Solitary palm. *Stem* to 15 (20) m tall, without leaf sheaths to 120 cm diam.; trunk dull brown, marked with broad, diamond-shaped leaf base scars. *Leaves* arching, 5 – 6 m long; leaf base 25 – 30 cm wide; pseudopetiole to one fifth of total leaf length; leaf sheath reddish-brown, fibrous; acanthophylls proximally congested in arrangement, pointing in several directions, green when young, becoming yellow, to c. 20 cm long, conspicuously folded (conduplicate); leaflets closely and regularly inserted in one plane of orientation, to c. 200 on each side of rachis, often forward-pointing, c. 25 – 30 cm long; lamina concolorous, bluish-green, with adaxial and abaxial surfaces glabrous. *Staminate inflorescence* erect; prophyll splitting twice between margins, yellow-green with reddish-brown tomentum when young becoming brown and coriaceous, to c. 40 cm; peduncle to c. 50 – 70 cm long. *Staminate flowers* crowded along full length of rachillae; calyx an even-rimmed cupule, 1.5 – 2 mm high; petals to 6 × 3 mm, with apex rounded and minutely serrate. *Pistillate inflorescence* initially erect, becoming pendulous; prophyll splitting between margins, yellow-green, to 60 × 10 cm; peduncle yellow-green, elongating with maturity, 1.6 – 2 m long; rachillae yellow, elongating with fruit maturation, to c. 60 cm long. *Pistillate flowers* mostly in distal half of rachillae, yellow-white, with faintly sweet scent; calyx cupule c. 2.5 mm high; petals c. 3 × 4 mm. *Fruit* obovoid, 1.5 – 2.0 × c. 1.2 cm, ripening from yellow-green to golden-yellow. *Seed* ovoid in shape, c. 15 × 10 mm, with rounded apices; embryo lateral opposite raphe; endosperm homogeneous.

DISTRIBUTION. *Phoenix canariensis* is endemic to the Canary Islands and occurs scattered, in populations of varying sizes, on all seven islands. The largest populations of wild palms are found on La Gomera.

HABITAT AND ECOLOGY. From sea-level up to 600 m in a range of habitats, from humid areas just below cloud forest to semi-arid areas where its presence usually indicates groundwater. Ecological requirements of *P. canariensis* were extensively studied by Lüpnitz & Kretschmar (1994). In its native habitat *P. canariensis* flowers during the spring and fruits ripen in the autumn.

SELECTED SPECIMENS EXAMINED. CANARY IS. La Gomera, 1895 (pist.), *Bourgeau* 1014 (BM!, FI-W!); Tenerife, Orotava, 2 July 1900 (pist.), *Bornmuller* 1265 (K!); Tenerife, 1933 (pist.), *Asplund* 858 (K!).

VERNACULAR NAMES. Palmera Canaria (Canary Islands), (Carlo Morici, *pers. comm.*).

USES. *Phoenix canariensis* is extensively cultivated in warm temperate regions as a street tree or garden plant. The leaflets are used in much the same way as those of *P. dactylifera* for a range of woven products including crosses for Palm Sunday celebrations. Inflorescence buds are tapped for the sweet sap which is eaten as palm honey. Mifsud (1995) reported an unusual use for leaves of *P. canariensis* in Malta where fishermen attract pilot and dolphin fish by floating two or three palm leaves on the sea surface near their nets. These fish species are known to congregate under floating objects and so are easy prey beneath the palm leaves.

CONSERVATION STATUS. The greatest threat to *P. canariensis* is an increase in cultivation of exotic species of *Phoenix* on the Canary Islands and contamination of the native species with alien genetic material. The ease with which species of *Phoenix* hybridize in cultivation is well known (Corner 1966; Hodel 1995), and the large number of horticultural names associated with 'canariensis-like' palms reflects the number and variety of hybrids in existence. *Phoenix dactylifera* and *P. roebelenii* have long been in cultivation on the Canary Islands and in recent years other exotic species of the genus have been introduced. Hybridization between *P. canariensis* and *P. dactylifera* poses the biggest problem due to the difficulty of early detection and removal of the resulting hybrids. The recent ban on the importation of exotic species of *Phoenix* should help lessen the hybridization threat. Importation of palms known to carry the pathogen that causes Lethal Yellowing may also pose a threat to wild populations of *P. canariensis.*

NOTES. In the classical literature a reference to *P. canariensis* was given by Pliny (see Hort 1916) as *Palmeta caryotas ferentia*, who reported the observation of Juba that '...Canaria also abounds in palm groves bearing dates.' Webb & Berthelot (1847) were the first to recognise differences between the Canary Island palm and the date palm, describing the former as *P. dactylifera* var. *jubae* Webb & Berthel. Christ (1885) later gave the palm species status as *P. jubae* (Webb & Berthel.) D. H. Christ. Neither name was in common use by the horticultural trade who tended rather to adopt the unpublished names *P. cycadifolia* and *P. canariensis* (e.g., Neubert 1873).

The name *P. cycadifolia* was validated by Regel (1879) with a brief description and illustration of a palm growing in Athens. The name *P. canariensis* was validated by Chabaud (1882) with a description and illustration of a cultivated palm grown from seed of Canary Island origin. Despite the robust nature of the '*canariensis*-like' palm depicted in the illustration of Regel (1879), Beccari (1890) considered *P. cycadifolia* to be a synonym of *P. dactylifera.* I agree with Moore (1971a) in considering the illustration of *P. cycadifolia* to be more similar to *P. canariensis* than *P. dactylifera*, and thus consider it a synonym of the former. The name *P. cycadifolia* predates *P. canariensis* by three years, should thus take nomenclatural precedence but because of the doubt surrounding the identity of *P. cycadifolia*, and the great familiarity of botanists and the horticultural trade with the name *P. canariensis*, I follow Moore (1963a) in maintaining the latter as the accepted name for the Canary Island palm.

7. Phoenix sylvestris *(L.) Roxb.*, Hort. Bengal.: 73 (1814) and Fl. Ind. ed. 2: 787 (1832); Royle, Ill. Bot. Himal. Mts.: 397, nom. (1840); Griff., Calcutta J. Nat. Hist. 5: 350 (1845); Mart., Hist. Nat. Palm. 3: 270, t. 136 (1849); Griff., Palms Brit. E. Ind.: 141, t. 228 (1850); Aitch., Cat. Pl. Punjab Sindh: 143 (1869); Brandis, Forest Fl. N.W. India: 554 (1874); Becc., Malesia 3: 364, t. 43, 3, f. 25 – 36 (1890); Becc. & Hook. f., Fl. Brit. India 6: 425 (1892); Gamble, Man. Ind. Timb.: 731 (1902); Brandis, Indian Trees: 645 (1906); L. H. Bailey, Stand. Cycl. Hort.: 2594 (1916); Blatt., Palms Brit. Ind.: 3, pl. 2, 3, (1926); Osmaston, Forest Fl. Kumaon: 544 (1927); C. Fischer, Fl. Madras 3: 1559 (1931); Kashyap, Lahore Fl.: 250 (1936); H. E. Moore, Principes 7 (4): 157 (1963); Mahab. & Parthasarathy, J. Bombay Nat. Hist. Soc. 60 (2): 374 (1963); H. G. Champion & S. K. Seth, A Revised Survey of Forest Types of India (1968); H. E. Moore & J. Dransf., Taxon 28 (1, 2/3): 67 (1979); K. M. Matthew, Mat. Fl. Tamilnadu Carnatic: 367 (1981) and Fl. Tamilnadu Carnatic 3: 1674 (1983); Malik in Nasir & Ali (eds.), Fl. W. Pakistan 153: 24 (1984); Noltie, Fl. Bhutan 3 (1): 416 (1990). Lectotype: *Katou-Indel* Rheede, Hort. Malab. 3: 15 – 16, pl. 22 – 25 (1682) vide Mart., Hist. Nat. Palm. 3: 270 – 273, t. 136 (1849). See Moore & Dransfield (1979).

Elate sylvestris L., Musa Cliff.: 11 (1736) and Sp. Pl.: 1189 (1753). Typified as for *P. sylvestris* above.

Katou Indel Rheede ex Buch.-Hàm., Trans. Linn. Soc. London 15: 82 – 87 (1826).

Solitary tree palm. *Stem* to 10 – 15 (20) m tall, without leaf sheaths c. 20 – 30 cm diam., with persistent, diamond-shaped leaf bases; stem base with mass of roots. Crown hemispherical, with more than 50 leaves. *Leaves* c. 1.5 × 4 m long; leaf sheath reddish-brown, fibrous; pseudopetiole 40 – 50 cm long × 3 – 5 cm wide at base; acanthophylls closely inserted, arranged in several planes, c. 13 – 18 on each side of rachis, conduplicate, yellow-green, very sharp, 4 – 14 cm long; leaflets irregularly fascicled, arranged in several planes, c. 80 – 90 on each side of rachis, concolorous, greyish-green, often waxy, very sharp, 18 – 35 × 1.2 – 2.4 cm. *Staminate inflorescences* to 25 per plant, erect, not extending far beyond prophyll; prophyll coriaceous, bright orange internally when young, splitting first adaxially (side adjacent to trunk), 25 – 40 × 6 – 15 cm; peduncle 20 – 30 × 1.2 – 2.2 cm; rachis 13 – 18 cm long with numerous, congestedly arranged rachillae, each 4 – 16 cm long. *Staminate flowers* white-yellow, musty-scented; calyx a deep cupule to 2 – 2.5 mm high with 3 poorly defined lobes; petals 3 (rarely 4), apices obtuse, slightly hooded, 6 – 10 × c. 3 mm; anthers 3 – 4 mm long. *Pistillate inflorescences* erect, arching on fruit maturation; peduncle green and upright, becoming golden-orange and arching on fruit maturation, to c. 90 × 2 cm; prophyll papery, short, splitting twice between margins, c. 24 × 5 cm; rachillae arranged in irregular horizontal whorls, c. 50 – 60 in number, yellow-green in colour, c. 8 – 34 cm long. *Pistillate flowers* creamy-white, c. 40 – 50 mostly restricted to distal half of rachilla; calyx cupule 1.5 – 2.5 mm high; petals 3 – 4 × 4 – 5 mm. *Fruit* obovoid, 15 – 25 × 12 mm, ripening from green to orange-yellow, with mesocarp moderately fleshy and astringent. *Seed* obovoid with rounded apices, 15 – 20 × 7 – 10 mm; embryo lateral opposite raphe; endosperm homogeneous.

DISTRIBUTION. *Phoenix sylvestris* is common, wild or cultivated, in the plains of India and Pakistan.

HABITAT AND ECOLOGY. *Phoenix sylvestris* thrives from the plains to the coast in low-lying wastelands, scrub forest and areas that have been disturbed or are prone to periodic or seasonal inundation with water, causing water-logging. In its native habitat *P. sylvestris* flowers at the beginning of the hot season from January to April, and fruits ripen from October to December.

SELECTED SPECIMENS EXAMINED. BANGLADESH. Chittagong, 10 Jan. 1851 (pist., stam.), *Hooker* 526 (K!); Chittagong (stam.), *Hooker* s.n. (K!). INDIA: BIHAR. Chota Nagpur, Feb. 1876 (stam., pist.), *Wood* s.n. (CAL!, DD!). PUNJAB. between Meerut and Delhi, 19 March 1962 (stam.), *Nair* 20818 (BSD!). RAJHASTHAN. Jhalawar, Batta rd., 28 May 1965 (pist.), *Wadhwa* 9519 (CAL!); Udaipur, Barman Hill, 4 Jan. 1986 (pist.), *Swami* 779 (BSD!). TAMIL NADU. Jaupore Distr., Negapatam, 25 July 1932 (pist.), *Jacob* s.n. (K!); Trichy Distr., Srirangam Is., 1 April 1976 (pist.), *Matthew* 1843 (RHT!); Trichy Fort Station, 30 Jan. 1995 (pist.) *Matthew & Barrow* 58, 59 (K!). WEST BENGAL. Dholutupur, Comilla, 5 Feb. 1943 (stam.), *Sinclair* 2840 (E!). UTTAR PRADESH. Moradabad, March 1844 (stam.), *Thomson* (K!). NEPAL. Kamali valley, btn. Manona and Badalkot, 25 April 1952 (pist.), *Polunin et al.* 3974 (BM!).

VERNACULAR NAMES. INDIA. Ita chettu (Telinga), [Beccari (1890)]; khurjjuri, kharjura, madhukshir (Sanscrit), khujjoor, kajar, kejur (Bengali), khaji, sendhu, kejur, khajur, khaji, salma, thalma, thakil (Hindi), ichal, kullu, ichalu mara (Kanara), khejuri (Uriya), itchumpannay, periaitcham, itcham-nar, itham pannay (Tamil), ita, pedda-ita, itanara, ishan-chedi (Telinga), [Blatter (1926)]; eechamaram, periya eecham (Tamil), [Matthew (1983)]; kubong, rotong (Lepchas), [Noltie (1994)]. PAKISTAN. Khaji, khajoor, [Malik (1984)]; taree-khajoor, [Aitchison (1869)].

USES. In parts of India, particularly West Bengal, sweet sap is tapped from the stem of *P. sylvestris* and drunk fresh or processed into a dark sugar (gur or jaggery) or alcoholic toddy (Davis 1972). The astringent fruits are rarely eaten fresh but are processed as jellies and jams. Blatter (1926) noted the fruits to comprise one constituent of a natural restorative, and the seeds when ground up with the root of *Achyranthes aspera* L. (*Amaranthaceae*) and chewed with betel leaves (*Areca catechu* L., *Palmae*) are considered a remedy for 'ague'.

CONSERVATION STATUS. Not threatened.

TAXONOMIC STATUS. *Phoenix sylvestris* was first described as *Katou-Indel* by Rheede (1678 – 1703) in *Hortus Indicus Malabaricus*, upon which Linnaeus' description of *Elate sylvestris* in *Musa Cliffortianus* (Linnaeus 1736) was entirely based. The description of *Elate sylvestris* in *Species Plantarum* (Linnaeus 1753), comprised two elements: *Palma dactylifera minor humilis sylvestris fructu minori, Hin Ind. Zeylaneus* of Hermann (1698) in *Paradisi Batavi Prodomus* 361, and *Palma sylvestris malabarica, folio acuto, fructu prunifacie* in *Historia Plantarum* 1364 (Ray 1686 – 1704). The latter was based entirely on Rheede's *Katou-indel*. Roxburgh (1832), in transferring *Elate sylvestris* to *Phoenix*, failed to acknowledge these two elements and based *Phoenix sylvestris* solely upon *Katou-indel*. The name *Phoenix sylvestris* is thus correctly typified by *Katou-indel* of Rheede's *Hortus Indicus Malabaricus*.

Hamilton (1827) recognised the two elements in *Elate* sylvestris but it was Martius (1823 – 53) in *Historia Naturalis Palmarum* who formally separated them. *Palma dactylifera minor humilis sylvestris fructu minore, Hin Ind. Zeylaneus* of Hermann was included by Martius in *P. pusilla* Gaertn., and *Katou-indel* was taken to refer only to *P. sylvestris* Roxb.

8. Phoenix caespitosa *Chiov.*, Fl. Somala 1: 317 (1929); Thulin, Fl. Somalia 4: 272 (1995). Lectotype: Somalia, 'Scorasar' valley, 1 July 1924 (pist.), *Puccioni & Stefanini* 672 (738) (FT!).

P. arabica Burret, Bot. Jahrb. Syst. 73 (2): 189 – 190 (1943). Type: Yemen, Wadi Madfar, Hadjeila, 700 m alt., Feb. 1889, *Schweinfurth* 993 (?B).

Stemless palm, clustering to form extensive thickets. *Leaves* to 3 m long; acanthophylls ± regularly arranged in one plane of orientation, to c. 13 on each side of rachis, to 8 cm long; leaflets ± irregularly arranged in one to two planes of orientation, often fascicled proximally, c. 20 – 50 on each side of rachis, 15 – 50 × 0.8 – 1.5 cm, narrowing to a sharp point, stiff, glaucous; lamina concolorous, drying pale green. *Staminate inflorescences* erect; prophyll coriaceous, splitting twice between margins, to c. 40 × 3 – 4 cm; peduncle to c. 50 × 1.2 cm; rachillae c. 60 in number, to

15 cm long. *Staminate flowers* with calyx cupule 2 mm high; petals 4 – 6 × 3 mm, with rounded apices. *Pistillate inflorescences* erect; prophyll not seen; peduncle to c. 40 cm long; rachillae c. 15 in number, occasionally branched to one order, to 24 cm long. *Pistillate flowers* with calyx cupule 2 – 3 mm high; petals 3 – 4 × 4 mm. *Fruit* sphaeroid-ovoid, 10 – 16 × 8 – 12 mm, ripening deep orange to purplish-brown, with fleshy, sweet, edible mesocarp. *Seed* ovoid, 12 × 8 × 8 mm; embryo lateral opposite raphe; endosperm homogeneous.

DISTRIBUTION. *Phoenix caespitosa* was described from Somalia (Chiovenda 1929) where it is found in the Sanaag, Nugaal and Bari regions of the north. Moore (1971b) noted a caespitose *Phoenix* taxon, growing with *Livistona carinensis* (Chiov.) J. Dransf. & N. W. Uhl, near Bankoualé in Djibouti, which he attributed to *P. caespitosa*. This report is not substantiated with a voucher specimen. Across the Gulf in the Arabian peninsula, Collenette (1985) recorded the species fairly widespread in the Asir and southern Hijaz of Saudi Arabia. The synonymous taxon, *P. arabica* Burret, was recorded from the Yemeni Arab Republic (Burret 1943).

SELECTED SPECIMENS EXAMINED. SOMALIA. Migiurtini coast, between 'Erèri Jellehò' and 'Martisor Dinsai', 31 May 1924 (ster.), *Puccioni & Stefanini* 659 (FT!); Hamur to Gombeia, 26 June 1924, *Puccioni & Stefanini* 926 (stam.), 927 (pist.) (FT!); Iskushuban, c. 10°16'N, 50°14'E, c. 280 m alt., 7 July 1980 (pist.), *Gillett* 23055 (K!); Halin Tug, 26 Oct. 1944 (pist.), *Glover & Gilliland* 236 (K!); 31 km N of Carin, c. 17 km S of Bosaso, 8 Jan. 1973 (stam.), *Bally & Melville* 15690 (K!); between Bosaso and Karin, 11°05'N, 49°10'E, 100 – 300 m alt., 12 Jan. 1995 (ster.), *Thulin et al.* 9016 (K!, UPS).

HABITAT AND ECOLOGY. In dry wadis, semi-desert bushland and rocky crevices and ravines up to 900 m in Somalia (Thulin 1995), up to 1950 m in Saudi Arabia (Collenette 1985). The species is often found in extensive thickets.

VERNACULAR NAMES. SOMALIA. Balmo, mayro/mairu (plant), awang (fruit) (Somali), [Abdi Shire (*pers. comm.*)]. YEMEN. Schottob (Hadjeih), Schegja (Ussil), [Burret (1943)].

USES. The fruits of *P. caespitosa* have a sweet and moist mesocarp layer, making them much sought after by animals including humans in Somalia (Abdi Shire, *pers. comm.*).

NOTES. *Phoenix caespitosa* is a dwarf palm described by Chiovenda (1929) from Somalia, and later recorded from Saudi Arabia and Yemeni Arab Republic. Of the many names published for *Phoenix* palms in Africa most refer to the polymorphic species *P. reclinata*. However, I consider *P. caespitosa* to be a distinct species. Staminate flower petals of *P. caespitosa* are rounded apically, as against the acute to acuminate petal apices of staminate flowers of *P. reclinata*.

9. Phoenix dactylifera L., Sp. Pl.: 1188 (1753) & Hort. Cliff.: 482 (1738); Willd., Linn. Sp. Pl. (ed. 4), 4 (2): 730 (1806); Roxb., Fl. Ind. ed. 2: 786 (1832); Kunth, Enum. Pl. 3: 255 (1841); Mart., Hist. Nat. Palm. 3: 257, t. 120, X. f. 1, t. 1., f. 1 (1849); Aitch., Cat. Pl. Punjab Sindh: 143 (1869); Boiss., Fl. Orient. 5: 47 (1882); Becc., Malesia 3: 355, tab. 43, f. 1 – 14 (1890); Hand.-Mazz., Ann. K. K. Naturhist. Hofmus. 28: 36 (1914); Brandis, Forest Fl. N.W. India: 552 (1874); L. H. Bailey, Stand. Cycl. Hort.: 2594 (1916); V. H. W. Dowson, Dep. Ar. Iraq Mem. 3 (3): 1 (1920), (2): 1

(1921), (3): 1 (1923) and Trop. Agric. 6, 6: 172 (1929); Blatt., Palms Brit. Ind.: 24, pl. 8, f. 3 (1926); Magalon, Contr. Étud. Palmiers Indoch.: 20 (1930); Vasc. & Franco, Portugaliae Acta Biol., Sér. B, Sist. 2: 312, figs. 3, 19-1 (1948); Blatt., J. Indian Bot. Soc. 11 (1): 42 (1932); Täckh. & Drah, Fl. Egypt: 165 – 273 (1950); H. E. Moore, Principes 7 (4): 157 (1963); P. Munier, Tech. Agric. Produc. Tropic. 14: 1 – 221 (1973); H. E. Moore, Fl. Iranica 146: 4 (1980); Malik, Fl. Pakistan 153: 23 (1984); J. Dransf., Fl. Iraq 8: 263 (1985). Lectotype: *Palma hortensis mas et foemina* Kaempf., Amoen. Exot. Fasc. 668, 686, t. 1, 2 (1712), vide Moore & Dransfield, Taxon 28: 64 (1979).

Palma dactylifera (L.) Mill., Gard. Dict. ed. 8, *nom. illegit.* (1768). See Moore (1963b). Typified as for *P. dactylifera* above.

Phoenix excelsior Cav., Icon. 2: 13, t. 125 (1793). No type designated.

P. dactylifera var. *cylindrocarpa* Mart., Hist. Nat. Palm. 3: 258 (1849). No type designated.

P. dactylifera var. *gonocarpa* Mart., Hist. Nat. Palm. 3: 258 (1849). No type designated.

P. dactylifera var. *oocarpa* Mart., Hist. Nat. Palm. 3: 258 (1849). No type designated.

P. dactylifera var. *oxysperma* Mart., Hist. Nat. Palm. 3: 258 (1849); Becc., Malesia 3: 357 (1890). No type designated.

P. dactylifera var. *sphaerocarpa* Mart., Hist. Nat. Palm. 3: 258 (1849). No type designated.

P. dactylifera var. *sphaerosperma* Mart., Hist. Nat. Palm. 3: 258 (1849). No type designated.

P. dactylifera var. *sylvestris* Mart., Hist. Nat. Palm. 3: 258 (1849). No type designated.

P. dactylifera var. *adunca* D. H. Christ ex Becc., Malesia 3: 357, pl. 43, f.12 (1890). Type: No type has been designated, but a specimen of three seeds (FI-B!) collected by Christ in 1888 has been annotated by Beccari with this name.

P. dactylifera var. *costata* Becc., Malesia 3: 357, pl. 43, f. 11 (1890). No type designated.

P. atlantica A. Chev. var. *maroccana* A. Chev. in Compt. Rend. Hebd. Séances Acad. Sci. t. 234, 2: 172 (1952) and Rev. Bot. Appliq. Agric. Trop. 32, no. 2311: 82 (1952). Type: No type was cited but Chevalier annotated a specimen of *P. dactylifera* in the Paris herbarium with this name (Morocco, between Tiznit and Agadir, 18 Dec. 1951 (pist.), *Chevalier* s.n.).

Solitary, or sparsely clustering palm, with several suckering offshoots at base. *Stem* to 30 m tall, without leaf sheaths to c. 40 – 50 cm diam.; trunk dull brown, marked with diamond-shaped leaf base scars c. 10 × 25 – 30 cm. *Leaves* straight, obliquely vertical in orientation, to 3 – 4 (5) m long; leaf base 15 – 20 cm wide; pseudopetiole 50 – 100 cm long; leaf sheath reddish-brown, to c. 45 cm long, fibrous; acanthophylls sparsely arranged, pointing in several directions, to 20 cm long; leaflets variously arranged in 1 – 3 planes of orientation, c. 50 – 130 on each side of rachis, stiff, c. 40 × 2 cm in length; lamina concolorous, glaucous, drying pale green. *Staminate inflorescences* erect; prophyll splitting 1 – 2 times between margins, yellow-green with reddish-brown tomentum when young, becoming brown and coriaceous, to 45 × 12 cm; peduncle to c. 50 cm long; rachillae to 30 cm long. *Staminate flowers* crowded along full length of rachillae; calyx a 3-lobed cupule with uneven margin, loosely surrounding the corolla; petals, 3 (rarely 4), creamy yellow-white, fleshy, each $7 - 10 \times 3 - 5$ mm with apex rounded and minutely serrate; stamen c. 5 mm. *Pistillate inflorescences* initially erect, becoming pendulous with maturity; prophyll splitting between margins, yellow-green, c. 100 cm long; peduncle yellow-green, 60 – 150 cm, greatly elongating after fertilisation; rachillae c. 150 in number, yellow, to c. 40 cm long, elongating with fruit maturation. *Pistillate flowers* mostly in distal half of rachillae, yellow-white, with faintly sweet scent; calyx cupule c. 2 – 3 mm high; petals, 3 (rarely 4), c. $4 - 5 \times 4$ mm. *Fruit* very variable in shape and size, $4 - 7 \times 2 - 3$ cm, ripening a range of colours from yellow and green to orange, red, purplish-brown to black; mesocarp sweet, thick and fleshy or dry and thin. *Seed* variable in size and shape but generally elongate, $20 - 30 \times 5 - 8$ mm, with apices rounded or pointed; embryo lateral opposite raphe; endosperm homogeneous.

DISTRIBUTION. The natural distribution of *P. dactylifera* is not known. The long history of date palm cultivation in the Middle East and North Africa has extended the distribution of the species far beyond its presumed original range, such that its area of origin remains a mystery. It is doubtful whether *P. dactylifera* still exists in the wild. Zohary & Spiegel-Roy (1975) claim that 'spontaneously-growing dates can be found throughout the range of date cultivation'. Many of these 'wild' date stands may represent long neglected palm groves or escapes from such groves. In some areas of the Near East date palms can be found occupying primary niches and could perhaps represent wild *P. dactylifera* (Zohary & Hopf 1988).

CULTIVATION. Traditional areas of date palm cultivation have included the Middle East, Near East, North Africa, parts of north western India and Pakistan (Malik 1984). More recently, date palm cultivation has been established on a commercial level in California.

HABITAT AND ECOLOGY. An old Arab proverb says of the date palm that 'its feet shall be in a stream of water, and its head in the furnace of heaven'. The ability of *P. dactylifera* to thrive in hot, dry conditions with little or no rain, as long as there is constant moisture about the roots for healthy growth and seed germination, have made it the classic symbol of the oasis. Throughout its distribution the date palm is taken as a reliable indicator of ground water in wadis, crevices and rocky ravines. In addition to its resistance to hot, arid atmospheres, the date palm shows remarkable tolerance of high salinity and water-logging. Despite resistance to water-logging, date palms are very vulnerable to excess rainfall and high humidity. Nixon (1951) noted that date fruits mature properly only if rainfall during the fruit maturation period (July to October) is less than 1.5 cm. Date palms are best adapted to tropical or sub-tropical conditions where the average daily maximum temperature is over 35°C and frost is very rare (Nixon 1951).

REPRODUCTIVE BIOLOGY. The date palm has long been thought to be wind-pollinated. However, there is evidence for both anemophily and entomophily in *P. dactylifera* and other species of the genus. The staminate inflorescences produce copious amounts of pollen, typical of anemophily. The grains lack a sticky pollen-coat and are at the lower end of the wind-borne size range. The pistillate flowers show less obvious adaptation to anemophily, lacking an extensive stigmatic surface for capturing wind-borne pollen. Furthermore, Uhl & Moore (1971) identified what could be interpreted as nectaries at the base of the ovary which could suggest entomophily. Many kinds of insects are frequent visitors to date palm inflorescences, but their role as pollinators has not been conclusively demonstrated. It seems that the pollination syndrome of wild date palms involves both anemophily and entomophily. Herrera (1989) reported that the only other European palm, *Chamaerops humilis* L., is also pollinated by a combination of insects and wind, and Henderson (1986) suggested that this is a common syndrome in palms.

Many animals are involved in dispersal of wild dates, as is the case with most palms (Zona & Henderson 1989). Ridley (1930) recorded the dispersal of dates by bats (*Rousettus aegyptiacus*). Several authors (e.g., Parrott 1980) have noted partially-eaten dates impaled upon the sharp acanthophylls of date palm leaves and have attributed it, circumstantially, to the action of the Great Grey Shrike (*Lanius excubitor*). Cowan (1984) suggested that it is the action of the wind rather than shrikes which is responsible. The most significant role in date palm dispersal has without doubt been played by man. Date fruits are generally easily stored and transported, and have therefore been an important component of the Middle Eastern diet, particularly for long journeys across the desert. The Phoenicians were not only early date palm cultivators but great travelling tradesmen and were certainly responsible in part for the early spread of date palms.

SELECTED SPECIMENS EXAMINED. ALGERIA. 1888 (pist.), *Christ* s.n. (FI-B!). BAHRAIN Is. without precise locality, 21 Feb. 1926 (ster.), *Fernandez* 2863 (K!). EGYPT. Abusir, near El Merq, 21 March 1924 (stam.), *Simpson* 1826 (K!); Nile, (pist.), *Hall* s.n. (K!). ISRAEL. Jericho, 1 April 1913 (stam.), *American Colony, Jerusalem* 6985 (K!). LIBYA. Kufra oases, 23 Aug. 1963 (seedlings), *Cambridge Expedition* 18 (K!). MOROCCO. Between Tiznit and Agadir, 18 Dec. 1951 (pist.), *Chevalier* s.n. (P!); High Atlas, near Marrakech, 450 m alt., 26 June 1971 (pist.), *Bocquet* 11071 (BM!). PAKISTAN. N Baluchistan, (ster.), s.n. (CAL!). SAUDI ARABIA. Wadi Hebron, 1835 (pist.), *Schimper* 250 (FI-W!); Near Jizan Dam, 12 Jan. 1980 (ster.), *Chaudhary* E421 (E!). SOCOTRA. Hagghiher Mts, Kishen, 12°35'N, 50°03'E, May 1967 (pist.), *Smith & Lavranos* s.n. (K!). SYRIA. Lake Tiberius, 1860, *Hooker & Hanbury* s.n. (K!). UNITED ARAB EMIRATES. Kalba oasis, 20 Feb. 1985 (stam.), *Western* 722 (E!). SPAIN. Murcia, April 1854 (stam., pist.), *Déséglise* 2314 (BM!); Barcelona, March 1913 (stam., pist.), *Sennen* 1807 (BM!).

VERNACULAR NAMES. The names listed here refer to *P. dactylifera* as a species. The serious student of date palm varieties and cultivars must look to Popenoe (1973) for a comprehensive list of vernacular names, and their meanings. ARABIA. Usteh-khurma (fruit), nukhal (leaves), (Arabic), [Beccari (1890)]. Egypt. Balah (date palm), (Egyptian), [Täckholm & Drar (1950)]. INDIA. Pind, chirwi, bagri (fresh dates), bela (dry dates), khajur, chuhara (leaves), gadda, galli (palm 'cabbage'), (Hindi); payr-etchum manam (leaves), (Tamil); kharjurapu chettu, perita chettu (leaves), (Telinga), [Beccari (1890)]. IRAQ. Nakhla/Nakhl, (date palm), tamr (fully ripe dates), rutab (fresh, edible but only half-ripe dates), kurjan, khurma (leaves), (Arabic), [Dransfield (1985)]; gutla-i-khaur, tukhm-i-khurma (fruit), (Arabic), [Beccari (1890)]. TURKEY. Khurma (date palm), (Kurdish, Turkish), [Dransfield (1985)].

USES. Commercially *P. dactylifera* is one of the most important species in the family, after *Cocos nucifera* L. (coconut) and *Elaeis guineensis* Jacq. (oil-palm). Date palms have been cultivated in the Middle East and northern Africa for at least 5,000 years (Zohary & Hopf 1988). For some communities practising subsistence agriculture, the date crop provides an essential subsidiary income. The primary use of date palms is, of course, their nutritious fruit which is eaten fresh, dried or processed as one of a wide-range of date products. Date seeds are used as cattle fodder (seeds ground up or soaked in water or sometimes sprouted first), or are occasionally ground as a coffee substitute or adulterant, or for ornamental purposes (as jewellery). Stems are tapped for the sweet sap (date 'honey') which can be drunk fresh, or processed as sugar or fermented into a highly intoxicating beverage, referred to as 'The Drink of Life' in cuneiform inscriptions of the ancient Egyptians (Täckholm & Drar 1950). Tapping interferes somewhat with fruit production, and the number of times a palm can be tapped is limited. In addition to the fruit, vegetative parts of the date palm are put to many and diverse uses including building materials (leaves, trunks), fencing (leaves, midribs), thatch (leaves), rope (leaf sheath, leaflet and midrib fibres), fuel (all vegetative parts, but especially leaf-bases); packaging, padding and protection (leaf sheath fibre). The terminal bud can be eaten as a sweet, tender vegetable, though rarely so because only non-productive palms would be felled for such a purpose. Cutting of the terminal bud leaves a cavity which fills with a thick, sweet refreshing fluid that is drunk fresh or fermented. The palm is important in several Christian, Jewish and Muslim festivals (Goor 1967; Nixon 1951; Popenoe 1924).

CONSERVATION STATUS. The conservation status of wild *P. dactylifera* is difficult to ascertain due to the continuing doubt as to whether it exists in that state. As a species, *P. dactylifera* cannot be considered threatened due to its extensive cultivation; however, positive conservation action may be necessary at the infraspecific level if diversity of date cultivars is to be maintained. Intrinsic within the hundreds of cultivars is a large reservoir of genetic diversity that has been the source of palms of varying vegetative and fruit characteristics for date palm growers through the ages. Recent years have seen a decrease in the number of varieties regularly propagated in cultivation. As with landraces and cultivars of all crops, active cultivation is vital to survival and a cultivar is soon lost for ever if it is not regularly propagated.

HUMAN SELECTION OF THE DATE PALM. The suckering habit of *P. dactylifera* makes it well-suited to vegetative propagation and domestication. Date palms produce offshoots at the trunk base allowing simple clonal propagation of chosen palms. Selection of productive palms in combination with artificial pollination techniques has led to the development and recognition of hundreds of date palm cultivars, differing in fruit characters such as size, texture, fleshiness, colour, taste, sweetness and storage quality.

INFRASPECIFIC NOMENCLATURE. Few of the hundreds of date palm cultivars have been formally described. Study of cultivars has been restricted to a local or regional level, so that there is inconsistent and differential use of vernacular and cultivar names between regions, and even palm groves. A proliferation of names can arise when date palms of identical stock are distributed to groves with varying microclimates, resulting in phenotypic variation (Brac de la Perrière 1988). Traditionally, date palm cultivars have been classified according to moisture content which, in turn, defines use. In general, fruits are divided into three classes: dry dates, which require high temperatures and sun levels for maturation, and are easily stored; semi-dry dates, which are more moist, and also store well; soft dates which must be eaten fresh and therefore are less commonly exported. Popenoe (1973) listed 1500 'varieties' cultivated to varying extents throughout the zone of date palm growth, and noted that if all named 'varieties' were to be collected together they would number several thousands. The term 'variety' has not been clearly defined in respect to the date palm, and was used by Popenoe (1973) to refer to what I consider to be cultivars. Martius (1823 – 1853) described seven varieties of *P. dactylifera*, and Beccari (1890) a further two, using only loosely defined fruit characters. Many cultivars are potentially referrable to any one of these names because of the brevity of the descriptions so that the names are useless to date grower and botanist alike.

Local studies aside, there have been no attempts to construct a workable, consistent infraspecific classification for *P. dactylifera*. The most comprehensive attempt at such a study is that of Popenoe (1973) who gave a concise overview of cultivar history, geography and origin, discussed the nomenclatural confusion that surrounds most varieties, listed 1500 names and described the most commercially important ones in greater detail. *Phoenix dactylifera* cultivar taxonomy has not been an aim of this monograph.

HISTORY AND ORIGIN OF THE DATE PALM. Together with the olive, grape and fig, date palms were amongst the first fruit crops domesticated in the Old World (Zohary & Speigel-Roy 1975). The cultural and religious significance of date palms, over a recorded 5000 year history, reflects their economic importance. The date palm was associated with the primitive Semitic goddess (Ishtar or Astarte) who symbolised the creative force of nature (Popenoe 1973). In Muslim tradition, God created the date palm from dust left over after Adam was created, and Arabs consequently know it as the 'Tree of Life' (Goor 1967), a name that well suits its multiplicity of uses.

The origin of the date palm has been much debated and suggested areas include desert northwestern Africa (Fischer 1882; de Candolle 1884), tropical North Africa (Schweinfurth 1873; Grisebach 1872), Arabia (Bonavia 1885), Babylonia (Hehn 1888), the Persian Gulf (Beccari 1890; Werth 1934; Popenoe 1973; Zohary & Hopf 1988), Southern Persia (Boissier 1882) and Western India (Hamilton 1827). To resolve this question a multi-disciplinary approach is advisable, combining evidence from botanical and ecological data with historical, cultural and archaeological information.

Feral date palms occur throughout the range of cultivated date, notably in the southern, warm and dry Near East as well as the northeastern Saharan and north Arabian deserts. It is difficult to ascertain whether these represent wild plants or merely secondary escapes from cultivated groves. Zohary & Hopf (1988) claimed that true 'wild dates', bearing small dry, hardly comestible fruits, are found in some areas of the Near East, growing in deep ravines, cliffs, and inaccessible gorges, often indicating ground water. Baluchistan, lowland Khuzistan, and the southern base of the Zagros Range facing the Persian Gulf, in particular, were identified by Werth

(1955) and Zohary & Hopf (1988) as areas supporting populations of 'wild dates'. Zohary (pers comm.) reported spontaneous 'wild dates' bordering the Dead Sea on both Israeli and Jordanian sides. These Dead Sea populations comprise equal numbers of staminate and pistillate palms, they are sexually-reproducing and occupy primary habitats, such as wet escarpments, gorges, springs and seepage areas, suggesting that they are wild. A paucity of morphological characters makes differentiation of wild from feral plants difficult. Molecular data from a wide range of domesticated, wild and feral date palms may offer new and useful information. Nuclear and chloroplast DNA regions sequenced in the current study show insufficient variation to be informative at the varietal level within *P. dactylifera*.

Beccari (1890) looked to ecology to solve questions of date palm origins. Date palms thrive in hot, dry conditions with little or no rain, as long as constant moisture about the roots is available. Wet soil is required for seed germination and hot sun for ripening of the fruit. In the words of Pliny (see Rackham 1945): 'It likes running water, and to drink all the year round, though it loves dry places'. Theophrastus in his *Enquiry into Plants* (370 – 285 BC, see Hort 1916) was an early observer of the remarkable resistance of date palms to salinity, 'wherever date palms grow abundantly, the soil is salt, both in Babylon, they say, where the tree is indigenous, in Libya, in Egypt and in Phoenicia'. Beccari (1890) understood the ecological characteristics of date palms as indicating that *P. dactylifera* originated in a hot, dry land where the ground was wet and saline. The lands adjoining the Arabian Gulf fit these specifications exactly.

Palaeobotanical data provides further support for a Near Eastern origin of the date palm. Solecki & Leroi-Gourhan (1961) found *Phoenix* pollen, comparable to that of *P. dactylifera*, in sediment samples taken from the Mousterian layer D in Shanidar Cave, northern Iraq, dating to thousands of years before the start of Neolithic agriculture, possibly suggesting the pre-agricultural existence of date palms in the Near East. The earliest definite signs of date palm cultivation were noted by Zohary & Hopf (1988) to appear in Chalcolithic Palestine around 3700 – 3500 BC, and there is early indication of a date crop from contemporary lower Mesopotamia. From the Bronze Age onwards, date cultivation has been well-established in warm areas of the Near East.

Historical and cultural information must also be considered. Corner (1966) noted that 'the origin of the date-palm is as insoluble as ever and will remain so until there are minds commensurate with the contributions that palms have made to civilisation'. Study of early date cultivators, particularly the Assyrians, Babylonians and Phoenicians, their history, trade routes, language and culture can offer useful and complementary information to botanical data. Historical records report date palm cultivation by the Sumerians as early as 3000 BC in present-day southern Iraq (Nixon 1951). Around the same time, the Phoenicians arrived in Phoenicia ('The Land of the Date'), the area covering present-day Lebanon and parts of Syria. It is not known from where the Phoenicians came but certain traditions suggest the Persian Gulf. The name *Phoenix*, coined by the Greeks, supports the sea-going Phoenicians as early cultivators and traders of date palms and fruit. The Phoenicians travelled far, establishing trading posts and settlements along their route west through the Mediterranean.

By identifying the Persian Gulf area as the home of the date palm, the natural distribution of *P. dactylifera* occurs at the western edge of that of *P. sylvestris*, the Indian date. All students of the genus have acknowledged the close resemblance of the two species (e.g., Beccari 1890; Corner 1966). Griffith (1845) found the two species to be indistinguishable, and Hamilton (1827) considered *P. sylvestris* to represent 'merely the wild plant of the same species with that which is cultivated in Arabia and Africa: but this culture has wonderfully improved the fruit'. Their close relationship is supported by morphological, anatomical and molecular data generated by the current study. Systematic analysis in this paper resolves *P. dactylifera*, *P. sylvestris* and *P. theophrasti* in one clade. *Phoenix sylvestris* differs from the date palm primarily in its solitary habit, its short pseudopetiole with congested, conspicuously folded acanthophylls, its shorter infructescence peduncle bearing smaller fruits that are scarcely fleshy and non-comestible, and its ability to withstand wet conditions. It is not certain whether human selection over thousands of years could result in the differences between the two species; however, on consideration of the selective restrictions of clonal cultivation, it would seem that major morphological and physiological changes and adaptations would not be encouraged. Corner (1966) found the strictly solitary habit of *P. sylvestris* and its occupation of a different ecological niche as strong evidence against it as sister to the date palm. Whatever the history of the relationship between *P. sylvestris* and *P. dactylifera*, their similarity is beyond doubt, and the identification of the Irano-Arabian area as the home of the date palm, alongside the western limit of *P. sylvestris*, comes as no surprise.

10. Phoenix theophrasti *Greuter*, Bauhinia 3: 243 – 250 (1967) & Mus. Genève ser. 2, 81: 14 – 16 (1968); O. Kirchner, Jahrb. Class. Philolog., Suppl. 3: 451 – 539 (1875); W. M. Leake, Travels in Northern Greece: 3 (1835); Langeron, Bull. Soc. Bot. France 74: 130 – 139 (1927); C. Barclay, Ann. Mus. Goulandris 2: 23 – 29 (1974); Franco, Fl. Europaea 5: 268 (1980); Anon., Kew, IUCN Threatened Plants Committee (1983); M. Boydak, Çevre Koruma 18: 20 – 21 (1983), Istanbul Üniv. Orman Fak. Derg., A 33-1: 73 – 92 (1983), Biol. Conservation 32: 129 – 135 (1985), Istanbul Üniv. Orman Fak. Derg., A 36-1: 1 – 13 (1986) & Principes 31 (2): 89 – 95 (1987); Turland *et al.*, Flora of the Cretan Area: 194, map 1728 (1993) & Chamaerops 11: 19 – 21 (1993); M. Boydak & S. Barrow, Principes 39 (3): 117 – 122 (1995). Type: Crete, Sitía Prov., near Vai, 2 Oct. 1966 (pist.), *Greuter* 7650 (holotype: hb. Greuter; isotypes: B, E!, G, GB, hb. mus. Goulandris, hb. Phitos, hb. Zaffran, K!, LD, M, W).

Clustering tree palm. *Stem* to 17 m tall, without leaf sheaths c. 50 cm in diam., with leaves persistent in upper trunk, otherwise with persistent, diamond-shaped leaf bases. *Leaves* obliquely vertical in orientation, c. 2 – 4 m long; leaf sheath fibrous, reddish-brown; pseudopetiole 50 – 70 cm long; acanthophylls irregularly arranged in more than one plane, to 10 on each side of rachis, yellow to orange-green; leaflets irregularly arranged in one to two planes of orientation, c. 65 – 100 on each side of rachis, stiff, to 50 × 2 cm; lamina concolorous, glaucous, surfaces often white with waxy coating. *Staminate inflorescences* erect; prophyll coriaceous, splitting twice between margins, c. 45 × 8 cm; peduncle to c. 40 cm long; rachillae to c. 10 cm long. *Staminate flowers* yellow-white, with strong musty scent; calyx cupule 2 – 3 mm high;

petals 3 (rarely 4), 8 × 3.5 mm; stamens 6 (rarely 7). *Pistillate inflorescences* erect arching slightly with fruit maturity; prophyll to 50 × 6 cm; peduncle elongating on fruit set, to c. 70 cm; rachillae to c. 80 in number, elongating on fruit set, to c. 50 cm long. *Pistillate flowers* yellow-white, with 3-lobed calyx cupule 2 – 2.5 mm high; petals 3 (rarely 4), 2 × 3 mm. *Fruit* oblong, c. 15 × 10 mm, green-yellow to brown, with sparse, mealy, sweet mesocarp. *Seed* with rounded apices, 11 – 13 × 6 – 7 mm; embryo lateral opposite raphe; endosperm homogeneous.

DISTRIBUTION. The species was first described from Vaí in Crete, and is now recorded from nine coastal localities on that island (Turland *et al.* 1993). Since 1982 (Boydak 1983, 1985, 1986, 1987; Boydak & Yaka 1983) the species has been recorded from the Datça Peninsula and Kumluca-Karaöz regions of south western Anatolia in Turkey from sea-level to 350 m. A third locality for *Phoenix* in Turkey was recorded by Boydak & Barrow (1995) from Gölköy near Bodrum. The history of this population and its identity are unclear but I consider it most likely to be referrable to *P. dactylifera*.

HABITAT AND ECOLOGY. *Phoenix theophrasti* is found in coastal areas, either on steep calcareous cliffs and rocks within a few metres of the sea, or somewhat inland along moist valley floors, stream banks and rocky gullies. Occurrence of the palms invariably indicates a water-source. Salt tolerance of the species enables it to survive combined pressures of exposure to coastal winds and sea water.

SELECTED SPECIMENS EXAMINED. CRETE. Sitía Prov., nr. Vaí, 2 Oct. 1966 (pist.), *Greuter* 7650 (Holotypes: hb. Greuter. Isotypes: B, E!, G, GB, hb. mus. Goulandris, hb. Phitos, hb. Zaffran, K!, LD, M, W); Pré Veli, 27 June 1967 (ster.), *Barclay* 272 (K!); Vaí, 3 April 1970 (pist.), *Synge* 19 (K!); Vaí, 9 April 1974 (pist.), *Canon & Canon* 4303 (BM!); Kissamos, Hrisos, Kalitissa, 27 April 1989 (pist., photo.), *Turland* 93 (BM!); NW Toplou Monastery, 60 m alt., 30 March 1990 (seed, photo.), *Turland et al.* 115 (BM!). TURKEY. Datça peninsula, (pist.), *Boydak* s.n. (K!); Hurmalibük village (K!), 22 April 1994 (pist.), *Barrow & Boydak* 37, 37A (K!), 22 April 1994 (stam.), *Barrow & Boydak* 38 (K!); Finike, Kumluca-Karaöz Bay, 24 April 1994 (pist.), *Barrow & Boydak* 41, 42 (K!).

VERNACULAR NAMES. Vaion (palm leaf), (Crete), [Barclay (1974)].

USES. In Crete leaves of *P. theophrasti* are used in Palm Sunday celebrations (Barclay 1974), just as leaves of *P. dactylifera* are used elsewhere.

NOTES. *Phoenix theophrasti*, the Cretan Date Palm, has been known in the Mediterranean since classical times when it was recorded by Theophrastus in *Enquiry into Plants* (370 – 285 BC, see Hort 1916) and Pliny in *Natural History* (see Rackham 1945). However, it was not until 1967 that the species was described formally by Greuter who named it in honour of the Greek botanist-philosopher. *Phoenix theophrasti* occupies a narrow ecological zone in coastal areas of southwestern Turkey and Crete, in habitats similar to those of feral dates: deep ravines, gorges, and water seepage areas. At present, the species is known only from Crete and southwestern Turkey, although Turland *et al.* (1993) reported '*P. theophrasti*-like' palms from the East Aegean islands of Kalimnos, Nisiros and Simi. *Phoenix dactylifera* and *P. theophrasti* are easily confused, particularly when sterile, and thus new records of *P. theophrasti* must be treated with caution.

Phoenix dactylifera and *P. theophrasti* are poorly differentiated and I consider the species status of *P. theophrasti* to be in doubt. Greuter (1967) considered the clustering habit of *P. theophrasti* to be a key character for differentiation of the species from *P. dactylifera*. Habit was earlier also referred to by Pliny who noted 'Some palms in Syria (referring to the area of modern Israel) and Egypt divide into two trunks, and in Crete even into three, and some even into five'. Similarly, Theophrastus (see Hort 1916) noted that '...they say that the palms in Crete more often than not have this double stem, and some of them have three stems; and that in Laporia one with five heads has been known.' Although *P. theophrasti* is a multiple-stemmed species, this character does not always distinguish it from *P. dactylifera*. In cultivation *P. dactylifera* exists as a single main stem with abundant basal suckers. These offshoots are generally removed for vegetative propagation purposes; however, if they are left to grow, a clump of many stems may arise from some date palm cultivars, just as with palms of *P. theophrasti*.

All characters used to differentiate *P. dactylifera* from other species in the genus must be considered in the context of the long history of *P. dactylifera* cultivation and human selection of certain morphological characteristics. Selection has focused particularly on such characters as peduncle length and fruit size, and thus they are not ideal for species differentiation.

Phoenix dactylifera and *P. theophrasti* are difficult to differentiate on the basis of morphological and anatomical data such that the specific status of the latter species is debatable. Furthermore, molecular data supports the two species as close sisters. An extensive range of morphological variation is exhibited by *P. dactylifera*, and it is likely that the morphological charateristics of *P. theophrasti* fall within this range. Although this study comes close to considering the two species conspecific there is still, in my opinion, insufficient data to support such a decision. This problematic issue can only be resolved by an extensive morphological survey of *P. dactylifera* across its geographical range, and the morphological characteristics of the feral date palms of the Persian Gulf, in particular, must be clarified. If *P. theophrasti* is found to be very similar to the Persian date palms then I would consider it undoubtedly a synonym of *P. dactylifera*. Even if this is indeed proven to be so, there may still be a case for maintaining the name *P. theophrasti* to refer to non-cultivated, feral populations of palms occupying wild habitats. Populations of palms in Turkey, Crete and areas of the Near East (and possibly elsewhere) would then be referrable to *P. theophrasti*.

11. Phoenix acaulis *Roxb.*, Hort. Bengal.: 73 (1814), Pl. Coromandel 3: 70, t. 273 (1820), Fl. Ind. ed. 2, 3: 783 (1832); Buch.-Ham., Trans. Linn. Soc. London 15: 87 (1826); Royle, Ill. Bot. Himal. Mts.: 397, nomen (1840); Kunth, Enum. Pl. 3: 257 (1841); Griff., Calcutta J. Nat. Hist. 5: 344 (1845); Mart., Hist. Nat. Palm. 3: 274 (1849); Griff., Palms Brit. E. Ind.: 137, t. 228 (1850); Brandis, Forest Fl. N.W. India: 555 (1874); Kurz, Forest Fl. Burma 2: 535 (1877); Mason, Burmah 2: 142 (1883); Becc., Malesia 3: 397, t. 44, 4. f. 51 – 57 (1890); Becc. & Hook. f., Fl. Brit. India 6: 426 (1892); Brandis, Indian Trees: 645 (1906); T. Cooke, Fl. Bombay 2: 802 (1907); L. H. Bailey, Stand. Cycl. Hort.: 2595 (1916); Blatt., Palms Brit. Ind.: 15 (1926);

Osmaston, Forest Fl. Kumaon: 545 (1927); P. C. Kanjilal, Forest Fl. Pilighit, Oudh, Gorakhapur & Bundelkhand: 382 (1933); H. E. Moore, Principes 7 (4): 157 (1963); B. D. Naithani, Fl. Chamoli 2: 667 (1985); P. C. Pant, Flora of Corbett National Park: 158 (1986); Noltie, Fl. Bhutan 3 (1): 235 (1994). Type: t. 273, Roxb. Pl. Coromandel 3 (1820). The specimen *Buchanan-Hamilton* 2199 (E!) collected in Rishikhund, (3 April 1811) is attributable to *P. acaulis* and could represent the type specimen. Roxburgh (1832) in describing the species as *P. acaulis* Buch. makes no specific reference to any Buchanan-Hamilton specimen.

P. acaulis var. *melanocarpa* Griff., Calcutta J. Nat. Hist. 5: 346 (1845) and Palms Brit. E. Ind.: 138, t. 227 (1850). No type designated. Griffiths used this name to describe a specimen sent to him by Colonel Ouseley, who reported that the local people consider it different from *P. acaulis*. I have not been able to trace any such specimen.

Acaulous palm; *stem* bulbous, to 10 cm high, densely covered with persistent leaf base stumps. *Leaves* 0.6 – 1.8 m long; leaf sheath reddish-brown, fibrous; rachis 0.3 – 1.5 m long × 1.5 – 2 cm in diam. at base; acanthophylls closely arranged in more than one plane, to 9 cm long; leaflets arranged in sub-opposite groups of 4s – 5s in more than one plane or orientation, c. 16 – 24 on each side of rachis, linear, 8 – 36 × 0.5 – 1.4 cm, flaccid, with strong marginal nerves; lamina concolorous, pale green. Inflorescences held at ground level. *Staminate inflorescences* not extending beyond prophyll; prophyll papery and splitting in many places, 13 × 2 cm; peduncle c. 7 × 0.6 cm; rachillae arranged in one whorl, 10 – 15 in number, c. 8 cm long. *Staminate flowers* not seen. *Pistillate inflorescences* not extending beyond prophyll; prophyll papery, c. 25 × 4 – 6 cm; peduncle c. 9 – 12 × 1.4 cm, not extending on fruit maturity; rachillae arranged in one compact whorl, 15 – 20 in number, 4 – 14 cm, drying striate, with differential maturation of fruit along rachillae. *Pistillate flowers* c. 5 – 20 per rachilla, congested in arrangement, each subtended by a distinct rachilla swelling (bractiform notch), 3 – 10 mm long; calyx cupule 3 mm high; petals 5 – 6 × 4 mm. *Fruit* obovoid, 12 – 18 × 8 mm, ripening from green with scarlet apices to blue-black, with mesocarp scarcely fleshy and stigmatic remains prominently pointed (1 – 2 mm long). *Seed* elongate in shape, 10 × 5 mm, with rounded apices; embryo lateral opposite raphe; endosperm homogeneous.

DISTRIBUTION. Sub-Himalayan belt of northern India and Nepal. Griffith (1845) and Kurz (1877) recorded the species in Myanmar, but I have seen no specimens to support this.

HABITAT AND ECOLOGY. Griffith (1845) noted that *P. acaulis* grows in clay soil on elevated plains north of the Ganges river. The species occurs in open forest, scrublands, savannahs and pine forest understorey at 400 – 1500 m. In India *P. acaulis* flowers in the cold season from November to January with fruits ripening from April to June.

SELECTED SPECIMENS EXAMINED. INDIA: ASSAM. Khasia Hills, Jan. 1886 (stam., pist.), *Mann* s.n. (FI-B!, K!). UTTAR PRADESH. Rishikhund, 3 April 1811 (stam., pist.), *Buchanan-Hamilton* 2199 (E!); between Saharanpur and Najiiabad in Bijnor, 10 March 1887 (pist.), *Duthie* s.n. (DD!, K!); Mailani, between Lakimpur and Kheri, 20

April 1964 (pist.), *Malhotra* 31498 (BSD!); Corbett National Park, Dubriya Chaur, 25 April 1971 (pist.), *Pant* 43751 (BSD!); Chamoli Distr., Nigol Valley, 1500 m alt., 17 Feb. 1979 (pist.), *Naithani* 63743 (BSD!). NEPAL. Kamali Valley, between Badalhot and Sika, 1350 m alt., 24 April 1952 (pist.), *Polunin et al.* 3967 (BM!, E!); Chitwan National Park, South Meghauli, 200 m alt., 7 March 1996 (pist.), *Watson* 9615 (E!).

VERNACULAR NAMES. INDIA. Khajuzi, Chota khajur (Uttar Pradesh, Siwalik hills), (S. Biswas, *pers. comm.*); khajuri, pind khajur, jangly khajur (Hindi), schap (Lepcha), chindi, hindi, jhari, sindi, juno (Kurku), pinn khajur, Boichand, Yita, [Blatter (1926)].

USES. In times of scarcity, the bitter stem pith of *P. acaulis* has been used as a sago substitute (Blatter 1926). The fruits are sweet and edible, though scarcely fleshy, and are commonly eaten by animals (Roxburgh 1832).

CONSERVATION STATUS. The current distribution and conservation status of *P. acaulis* is unclear. During fieldwork in India for this study I failed to find and observe *P. acaulis* in its native habitat in Uttar Pradesh, despite visits to previous collection localities. However, the species has been reported, and recently collected, from Chitwan National Park in central Nepal, near the Indian border, and Dhar (1998) discusses the recent discovery of a population in Uttar Pradesh. It is assumed that, with the decline in forest habitats formerly ranging across the sub-Himalayan belt of northern India, the natural habitat for *P. acaulis* has suffered and the species is now restricted to more remote and inaccessible areas. A further threat to the species in some parts of its range has been the destructive harvest of the stem pith as a sago substitute. The bitter pith of *P. acaulis* was used heavily in India during a severe famine in the 1930's (Blatter 1926).

NOTES. There has been widespread misapplication of the name *P. acaulis*, as a result of confusion of the species with stemless individuals of *P. loureiri*. The presence of a stem can be a misleading character in the genus because individuals of several species remain stemless for extensive periods before a trunk develops. Environmental factors have an important role to play; it is common for populations of *P. loureiri* in dry or disturbed areas to consist only of stemless, shrubby individuals, whereas populations occurring in less stressful conditions consist of palms with well-developed trunks. Differentiation between the two species is especially difficult when sterile. Even when in staminate flower, the prophylls of both species split to reveal inflorescences at ground level. The difference between the taxa becomes apparent on fruit set: the fruiting peduncle of *P. loureiri* elongating greatly to present mature fruit beyond the leaves, whereas that of *P. acaulis* remains at ground level, nested amongst the leaf bases (Noltie 1994). Pistillate rachillae of *P. acaulis* are distinctly shorter, thicker and more congested with larger fruit than those of *P. loureiri*. Each fruit of *P. acaulis* is subtended by a thickening of the rachilla, referred to by Roxburgh (1832) as a 'bractiform notch'.

12. Phoenix pusilla *Gaertn.*, Fruct. Sem. Pl. 1: 24, tab. 9 (1788). Lectotype: Gaertner's illustrations of fruit and seed (tab. 9) in *De Fructibus et Seminibus plantarum* (1788).

Elate sylvestris L., Sp. Pl.: 1188 (1753), Sri Lankan plant only (referred to as *Hinindi*).

P. farinifera Roxb., Pl. Coromandel 1: 74, t. 74 (1796) & Fl. Ind. 2: 785 (1832).
   Willd., Linn. Sp. Pl. (ed. 4), 4 (2): 731 (1806); Griff., Calcutta J. Nat. Hist. 5: 348
   (1845); Becc., Malesia 3: 402, t. 44, fig. 3 (1890); T. A. Davis, & A. F. Joel, Palms &
   Cycads 23: 2 – 10 (1989); de Zoysa, Fl. Ceylon (in press). Type: t. 74 in Roxb. Pl.
   Coromandel 1 (1796); Roxburgh's illustration shows a pistillate inflorescence, a
   single staminate rachilla and illustrations of flowers of both sexes and a mature
   fruit. De Zoysa (in press) unnecessarily lectotypifies the name with the
   illustrations of Beccari in Malesia 3: 402, tab. 44, fig. 3 (1890).

P. zeylanica Trimen, J. Bot. 23: 267 (1885); Becc. & Hook. f., Fl. Brit. Ind. 6: 425
   (1892); Trimen, Handb. Fl. Ceylon 4: 326, pl. 95 (1898); Blatter, Palm. Brit. Ind.
   11: 14 (1926); Mahabalé & Parthasarathy, J. Bombay Nat. Hist. Soc. 60: 375
   (1963); De Zoysa, Fl. Ceylon (in press). Type: Sri Lanka (stam., pist.), *Thwaites*
   C.P. 3172 (K!, holotype; CAL!, PDA!, LEN, FI-B!, isotypes).

Solitary or clustering palm. *Stem* to 6 m high and 30 cm in diam. *Leaves* to 3 m long;
pseudopetiole to 70 cm long × 1.5 – 3 cm wide at base, rounded abaxially; leaf sheath
fibrous, reddish-brown; leaf bases persistent, vertically orientated on trunk, c. 8 cm
wide at base; acanthophylls individually arranged in one or more planes of
orientation, c. 7 – 18 on each side of rachis, yellow-green, very sharp, to 11 cm long;
leaflets more or less irregularly arranged, quadrifarious proximally, c. 30 – 100 on
each side of rachis, elongate-spathulate in shape with very sharp, needle-like apices, 10
– 45 × 0.5 – 3 cm in length; leaflet join with rachis marked by yellow-orange pulvinus;
lamina concolorous, dark, glossy green, and pliable in texture. *Staminate inflorescences*
erect; prophyll coriaceous, 12 – 30 × 4 – 8 cm; peduncle 5 – 25 cm long; rachillae
arranged at wide angle to the rachis, c. 50 – 70 in number, to 21 cm long. *Staminate
flowers* ovoid, yellow-white; calyx 1 – 1.5 mm high; petals 4 – 5 × 2 – 3 mm ovate, with
rounded apices. *Pistillate inflorescences* erect, arching at fruit maturity; prophyll
coriaceous, splitting twice, 17 – 41 × 2.5 – 5.5 cm; peduncle to c. 25 – 75 cm; rachillae
20 – 120 in number, orange-green, 4 – 30 cm long. *Pistillate flowers* mostly in the distal
half of rachilla; calyx to 1.2 mm high; petals 2 × 3 – 4 mm. *Fruit* ovoid, 11 – 15 × 5 – 8
mm, ripening from green to red to purple-black, moderately fleshy, sweet. *Seed* ovoid
with rounded apices, pinkish-brown when fresh, drying glossy chestnut-brown, 8 – 12
× 6 mm, with intrusion of testa in region of raphe (postament) often Y-shaped in
transverse section; embryo lateral opposite raphe; endosperm homogeneous.

DISTRIBUTION. Sri Lanka and Eastern Ghats of Tamil Nadu and southern region
of Kerala in India.

HABITAT AND ECOLOGY. Roxburgh (1832) described *P. farinifera* (= *P. pusilla*) as
a 'native of dry, barren ground, chiefly of the sandy lands at a small distance from
the sea near Coringa (Coromandel coast of south eastern India)'. However, *P.
pusilla* is not restricted to coastal areas in India but is also found inland at the
margins of marshes and raised banks along borders of paddy fields, up to 700 m
altitude. In Sri Lanka, *P. pusilla* is found in the dry lowlands of the north and east
(where it was previously referred to as *P. farinifera* Roxb.), and wetter lowlands and
hill country of the south west up to 500 m altitude (where it was previously referred
to as *P. zeylanica* Trimen).

SELECTED SPECIMENS EXAMINED. INDIA: ANDHRA PRADESH. Cuddapah Distr., Feb. 1883 (ster.), *Gamble* 11105, 11106 (K!). KERALA. Trivandrum, Veli, sea-level, 25 Jan. 1995, *Barrow & Bunoi* 51 (pist.), 52 (stam.) (K!, TBGRI!). TAMIL NADU. Mariappa nagar, 10 Jan. 1978 (stam.), *Mohan* 11282 (RHT!). (Coimbatore Distr.), 1882 (stam.), *Brandis* s.n. (FI-B!). (Madras Distr.): Chingleput, near Sadras, 23 Feb. 1933 (pist.), *Chevian Jacob* 80414 (K!). (Salem Distr.): (Mettur range), Peria Thanda, N Bagpur R.F., 19 Dec. 1976 (pist.), *Matthew & Alphonse* 5884 (RHT!). (Shevaroys South Range), Thekkampatti, 10 May 1978 (pist.), *Mohan* 13476 (RHT!). (Attur Range), Periakalrayans, 22 Sept. 1978 (pist.), *Venugopal & Manoharan* 17610 (RHT!). (South Arcot Distr.): Takarai R.F., 19 Jan. 1978 (pist.), *Ramamarthy* 52846 (CAL!); (Uludurpet Range), Pulloorkkadu, 3 Dec. 1979 (pist.), *Matthew* 25097 (RHT!); Vridachalam, 1 Feb. 1980 (pist.), *Matthew* 26255 (RHT!). (Tinnevelly Distr.), Nateriakal, c. 900 – 1200 m alt., 12 Feb. 1913 (ster.), *Hooper & Ramaswami* 38454 (CAL!). (Tanjore Distr.), Sobanapurum, 21 March 1978 (pist.), *Matthew* 12568 (RHT!); Srirangam Is, 31 Jan. 1995 (pist.), *Matthew & Barrow* 65A (pist.), 65B (stam.) (K!, RHT!). SRI LANKA. (Matale Distr.): Dambulla to Sigirya road, near Kibissa, 180 m alt., 8 Feb. 1995 (pist.), *Barrow & Weerasooriya* 69 (pist.), 70 (stam.), 71 (pist.) (K!, PDA!). (Galle Distr.): Kanneliya Forest Reserve, Hiniduma, 6 March 1992 (pist.), *De Zoysa* 11 (PDA!). (Jaffna Distr.): Point Pedro, Kaddaikadu, 16 March 1973 (stam.), *Bernardi* 14250 (PDA!). (Kurunegala Distr.): Wariyapola, 27 March 1970 (pist.), *Amaratunga* 2045 (PDA!). (Puttalam Distr.): Miriswatte, Negombo-Maradagahamula road, 1 Oct. 1992 (stam., pist.), *De Zoysa* 72 (PDA!). (Kalutara Distr.): Habungala, Bentota, 10 March 1972 (pist.), *Amaratunga* 2389 (PDA!). (Ratnapura Distr.): Sinharaja Forest, 8 March 1992 (stam.), *De Zoysa* 28 (PDA!). (Puttalam Distr.): Chilaw to Puttalam road, 3 Feb. 1995, *Barrow & Weerasooriya* 66 (pist.), 67 (stam.), 68 (stam.) (K!, PDA!).

VERNACULAR NAMES. INDIA. Chilta-eita (Telinga) [Roxburgh (1832)]; eethie (Tamil); chiruta-itu (Telinga); eentha (Malayam), [Blatter (1926)]; eecha maram (Tamil), [K.M. Matthew, *pers. comm.*]. SRI LANKA. Indi (Sinhalese); inchu (Tamil) [De Zoysa (in press)].

USES. Leaflets of P. *pusilla*, once stripped of the midrib, boiled and sundried, are used for various woven products in south and west Sri Lanka (De Zoysa, in press). The sweet fruits are often eaten by children.

CONSERVATION STATUS. Not threatened.

NOTES ON TAXONOMIC HISTORY OF *Phoenix* IN SRI LANKA. The first references to *Phoenix* in Sri Lanka are given by Hermann (1687, 1698). Hermann (1687, 1717) described two *Phoenix* species from Sri Lanka which were referred to as *Indi Hinindi* and *Indi Mahaindi*. From the brief descriptions it is clear that Hermann distinguished these palms on size. The vernacular names cited as *Maha Indi* and *Hin Indi* support this. In Sinhala, 'Indi' means 'date', 'maha' and 'hin' mean 'large' and 'small' respectively.

Linnaeus (1747) included the two taxa in separate genera when he cited *Indi Mahaindi* in *Phoenix* and *Hinindi* in *Vaga* L. In *Species Plantarum*, Linnaeus (1753) included *Mahaindi* in *Phoenix* and *Hinindi* in the genus *Elate* L. *Elate* also included *Katou-indel* of Rheede (1678 – 1703), which is attributable to *Phoenix sylvestris* of India. By including *Hinindi* and *Katou-indel* together under *Elate*, it would appear that Linnaeus was confused on two points. Firstly, he took the name *Hinindi* to refer to the larger of the two Sri Lankan palms, and secondly that the larger palm was synonymous with *P. sylvestris.* This confusion was repeated in later treatments of *Phoenix* in India and Sri Lanka (Martius 1823 – 1853; Thwaites 1864), but Hamilton (1827) correctly referred *Hinindi* to *P. farinifera.*

The first post-Linnean species of the genus in Sri Lanka was described as *P. pusilla* by Gaertner in 1788. Gaertner described the species from a short palm, said to have been of Sri Lankan origin, cultivated in Leiden Botanic Garden. The brief description and illustrations of fruit and seed are not sufficient for certain identification, and it is not clear whether the species refers to *Mahaindi* or *Hinindi.* However, the citation of the species as a native of Sri Lanka and East India suggests that *P. pusilla* refers to the smaller palm, *Hinindi.*

In 1795, in *Plants of the Coromandel Coast,* Roxburgh described *P. farinifera* as a short palm of sandy, coastal regions along the Coromandel coast of south eastern India but made no mention of Sri Lanka. In *Flora Indica,* Roxburgh (1832) acknowledged its existence in Sri Lanka, and cited *P. pusilla* Gaertn. as a synonym. No mention was made of *Mahaindi.* Griffith (1845) followed Roxburgh (1832) in his treatment of the genus in India.

Thwaites (1864) recorded only one species of *Phoenix* in Sri Lanka, referring it to *P. sylvestris* and noting it to be a native of hotter parts of the island. The specimen *C.P. 3172* is cited. It is clear that Thwaites (1864) referred to the tall-stemmed palm of south western Sri Lanka but, following Linnaeus, called it *Hinindi* rather than *Mahaindi.* The confusion between the tall-stemmed Sri Lankan *Phoenix* and *P. sylvestris* of India was appreciated by Trimen (1885) who described the Sri Lankan species as *P. zeylanica* Trimen. Trimen (1885, 1898), Blatter (1926) and Mahabalé & Parthasarathy (1963) all considered the smaller Sri Lankan *Phoenix* to be identical with *P. farinifera* of India, for which the name *P. pusilla* took precedence.

Only von Martius (1823 – 1853) and Beccari (1890) considered Gaertner's *P. pusilla* to refer to *P. zeylanica* rather than *P. farinifera.* Von Martius (1823 – 1853) cited two species of *Phoenix* in Sri Lanka as *P. pusilla* and *P. farinifera. Hinindi* is cited under *P. pusilla,* and no mention is made of *Mahaindi.* Beccari (1890) adopted this opinion after seeing Gaertner's illustration of seeds of *P. pusilla* which show the intrusion of the testa in the region of the raphe to be Y-shaped in transverse section. This illustration matched Beccari's own observations of seeds of *P. zeylanica,* and I have also found there is a tendency for the seeds of *P. pusilla* to show this character. However, the pattern of intrusion of the testa is generally too variable both within and between species and cannot be considered taxonomically reliable, as Beccari (Beccari & Hooker 1892 – 93) himself later acknowledged. Beccari & Hooker (1892 – 93) reversed this decision in Flora of British India, which cited *P. pusilla* as a synonym of *P. zeylanica.* De Zoysa (in press) considered Gaertner's description of *P. pusilla* to be inadequate for certain identification and recommended that the later names, *P. zeylanica* (Trimen 1885) and *P. farinifera* (Roxburgh 1832), be adopted.

Nomenclatural confusion surrounding species of *Phoenix* in Sri Lanka and southern India is, in part, a reflection of the poor delimitation of the taxa involved. Inadequate attention has been paid both to the relationship between Sri Lankan *Phoenix* palms and their Indian counterparts, and to the relationship between ecological and morphological variation. Greater consideration of these factors, in the context of variation within the genus as a whole, has clarified delimitation of *Phoenix* species and associated nomenclatural problems in Sri Lanka and southern India. For the reasons outlined below, I consider *P. zeylanica* and *P. farinfera* to be conspecific. Both names are predated by *P. pusilla* Gaertn. which therefore takes nomenclatural precedence.

NOTES ON SPECIES DELIMITATION. De Zoysa (in press) acknowledged the difficulty of clearly delimiting two *Phoenix* taxa in Sri Lanka, suggesting ecological factors as the cause, but chose to maintain them as distinct species. My observations of populations of *Phoenix* in southern India and Sri Lanka lead me to conclude that *P. farinifera* and *P. zeylanica* cannot be considered distinct. The key character of stem height is particularly unreliable because individuals of *P. zeylanica* can remain stemless for many years and *P. farinifera* can sometimes be found with a well-developed stem. Leaflet orientation is also unreliable as a distinguishing character. Clear distinction can be made between leaflets arranged in one or more than one plane, but distinction is less clear between those arranged in three or four planes. The number of planes of orientation appears to depend in part on leaf size and position. Proximal leaflets of *P. zeylanica* are strongly quadrifarious in arrangement, but less so distally. Leaflets of smaller individuals of *P. farinifera* are arranged in more than one plane of orientation, but less strongly four-ranked. De Zoysa (in press) describes the leaflet apices of *P. farinifera* and *P. zeylanica* to be 'softish' and 'very sharp' respectively. In my experience, leaflets of both taxa are sharply pointed, the apices marked by an almost needle-like extension.

To summarise, I consider the characters previously used to distinguish *P. farinifera* and *P. zeylanica* are insufficient for reliable species delimitation and therefore consider the taxa to be conspecific under the name *P. pusilla* Gaertn. Polymorphism within *P. pusilla* is likely to be due to ecological factors.

13. Phoenix loureiri *Kunth*, Enum. Pl. 3: 257 (1841) as '*Phoenix loureirii*'; H. E. Moore, Principes 7 (4): 157, 179 (1963), '*Phoenix loureirii*'. Lectotype: *Pierre* 4832 (FB-I!), collected by Harmand from Mount Kuang Repen in Cambodia.

Solitary or clustering palm. *Stem* to 1 – 4 (5) m, without leaf sheaths to c. 10 – 30 (40) cm in diam., with crowded diamond-shaped, persistent leaf-bases, internodes very short. *Leaves* to 2 m long; pseudopetiole 20 – 40 cm long; leaf sheath reddish-brown, fibrous; acanthophylls c. 15 on each side of rachis, yellow-green to orange, to 20 cm long; leaflets arranged in more than one plane of orientation, proximally fascicled in 3s – 4s, more regularly arranged apically, to 130 on each side of rachis, 20 – 45 × 0.5 – 2.3 cm, flaccid or stiff; lamina either concolorous or abaxial surface bluish-green with tannin in patches along midrib and continuous along leaflet margin. *Staminate inflorescences* erect; prophyll yellow-green in colour, to 40 × 7 cm (often much smaller); peduncle to c. 15 cm long; rachillae congested on rachis, c. 10 cm long. *Staminate flowers* sweet-scented initially, turning musty, with calyx a three-pointed cupule 1.5 – 2 mm high; petals yellow-white, oblong in shape, c. 4 – 6 × 2 – 2.5 mm, with apex roughly undulate and often thickened; anthers yellow-white. *Pistillate inflorescences* erect, often arching with fruit maturity; prophyll papery to coriaceous, splitting twice either along or between margins, c. 20 × 3 cm; peduncle to 1.5 m long; rachillae up to 40 in number, to 40 cm long, elongating with fruit set. *Pistillate flowers* with calyx cupule 1.5 – 2 mm high, yellow; petals orange-pink to yellow, 2 – 2.5 × 3 – 4 mm. *Fruit* restricted to the distal half to two thirds of rachilla, ovoid to obovoid, 9 – 18 × 5 – 9 mm, maturing from green to blue-black when ripe, with mesocarp mealy and slightly sweet. *Seed* obovoid, 11 – 18 × 6 – 9 mm, with rounded ends, and raphe extending full length of seed; embryo lateral opposite raphe; endosperm homogeneous.

DISTRIBUTION. Sub-Himalayan belt southwards through India, and eastwards through Indochina to southern China (including the islands of Hong Kong and Macao), Taiwan and to Batanes and Sabtang Islands of the Philippines.

HABITAT AND ECOLOGY. A variety of habitats from sea-level to 1700 m, in open scrublands or as part of the undergrowth of dry dipterocarp, mixed deciduous or pine forest. The species is very common in disturbed, anthropogenic areas such as seasonally-burnt grasslands, along roadsides or raised ground bordering rice-paddy. As with all species of the genus, staminate flowers of *P. loureiri* are visited by a range of insects, particularly beetles, but it is not known which are the pollinators and which are merely pollen thieves. The fruits are eaten by a range of birds and mammals attracted by the sweet but thin mesocarp.

USES. The leaflets of *P. loureiri* have many domestic uses, such as the manufacture of mats and brooms. In the Philippines shredded, sun-dried juvenile leaves are woven as raincoats (Gruezo & Fernando 1985). The apical bud is sweet and can be eaten as a vegetable (palm cabbage). The fruits are sweet, if a little mealy, and are commonly eaten by children. Padmanabhan & Sudhersan (1988) noted the medicinal use of the species by tribal people in southern India.

CONSERVATION STATUS. *Phoenix loureiri* is not considered to be threatened because it thrives in disturbed, anthropogenic areas. However, populations are declining in certain parts of their ranges. Padmanabhan & Sudhersan (1988) noted 'mass destruction' of the species on hillsides in southern India due to heavy pressure from harvesting for leaves. Gruezo & Fernando (1985) called for continued protection of the species in the Philippines, where it is found only in localised populations on Sabtang and Batanes Islands.

NOTES. The group of *Phoenix* palms from Asia that have been variously known as *P. humilis*, *P. hanceana* and *P. loureiri* are referred to here as the '*P. loureiri* complex'. Confusion has surrounded the taxonomy of the complex, which has long been poorly understood. Two taxonomic approaches can be adopted. The complex can be treated as one polymorphic species, or an attempt can be made to recognise distinct taxa within the complex with formal taxonomic status conferred upon them. The former approach was adopted by Moore (1963a) who recognised the complex as one wide-ranging species, *P. loureiri*. The latter approach was taken by Beccari (1890) in his monograph of the genus. He treated the '*P. loureiri* complex' as one species, *P. humilis*, comprising five closely-related varieties ranging from India to the Far East. Beccari (1890) acknowledged that varieties of *P. humilis* were not supported by stable taxonomic characters, and refers to them as 'regional forms'. Beccari (1890) thought *P. humilis* var. *hanceana* the most distinct variety, and later gave it species status as *P. hanceana*, itself comprising three varieties from southern China and Hong Kong, Taiwan and the Philippines (Beccari 1908).

An attempt has been made here to find support for distinct taxa within the '*P. loureiri* complex' using a combination of field, herbarium, anatomical and molecular data. Field data supports Beccari's (1890) observation that the '*P. loureiri* complex' is very adaptable and occurs in a wide variety of climates and habitats. It is this adaptability which may be the cause of polymorphism in the group. Molecular data group all members of the complex together and do not support any division of the complex into distinct taxa. Herbarium and anatomical studies have provided data which split the complex into two groups. Each group comprises a number of *P. humilis* varieties as defined by Beccari (1890, 1908). The taxa referred to by Beccari (1890) as *P. humilis* var. *loureiri* and var. *hanceana* of Indochina and the Far East respectively, have continuous strips of sclerotic, tannin-filled cells along the leaflet margins and discontinuous patches of such cells in the abaxial midrib region. In contrast, the Indian varieties of Beccari (1890), *P. humilis* var. *typica*, var. *pedunculata* and var. *robusta*, lack sclerotic, tannin-filled cells. Although the five *P. humilis* varieties (Beccari 1890) have been found to reflect certain geographical trends in morphological variation within the complex, I do not consider it possible to maintain them as distinct taxa.

I recognise the '*P. loureiri* complex' as one species, *P. loureiri* Kunth comprising two polymorphic varieties, var. *loureiri* and var. *humilis* (Becc.) S. Barrow. The nature and pattern of distribution of polymorphic characters is poorly known and it is unclear to what extent ecological factors play a role in determining intravarietal variation. More extensive sampling from a wide range of populations is required if further taxa are to be defined within the varieties of *P. loureiri*. However, the description of such intravarietal taxa would have ramifications on taxon delimitation elsewhere in the genus. It is my view that the polymorphism observed within the varieties of *P. loureiri* Kunth is equivalent to that included within the polymorphic African species, *P. reclinata*, within which I have not formally recognised infraspecific taxa.

var. **loureiri**

P. humilis var. *hanceana* Becc., Malesia: 379, 392 (1890). Type: Beccari (1890) notes
　　the specimen from Hong Kong, 1853 – 56 (pist.), *Wright* 507 (K!, LE).
P. hanceana var. *formosana* Becc., Philipp. J. Sci. 3: 339 (1908); Hui-Lin, Fl. Taiwan 5:
　　791, pl. 1524, 1525 (1978). No type designated.
P. hanceana var. *philippinensis* Becc., Philipp. J. Sci. 3: 339 – 342. Type: Batanes Is.,
　　Sabtang Island, June 1907 (pist.), *Fénix* (Bur. Sci.) 3744. (FI-B!).
P. pusilla Lour. (non Gaertn.), Fl. Cochinchina: 614 (1790), *nom. illegit.* No type
designated.

Distinguished from *P. loureiri* var. *humilis* by the presence of a continuous strip of sclerotic, tannin-filled cells along leaflet margin and discontinuous patches of such cells in the abaxial midrib region.

HABITAT AND ECOLOGY. In Thailand *P. loureiri* var. *loureiri* is found extensively on grassy or rocky slopes or in open areas under dry dipterocarp forest up to 500 m altitude, often but by no means exclusively on limestone soil. At higher altitude it is found on sandy soil in pine forest undergrowth up to 1700 m. The species is increasingly common, almost weedy, in the undergrowth of pine plantations. *Phoenix loureiri* var. *loureiri* is rare in Peninsular Thailand. Populations are known only from Trang province near the Malaysian border, growing in grasslands with intense human access, often on old termite mounds along the borders of neglected rice-paddy.

SELECTED SPECIMENS EXAMINED. CAMBODIA. Mount Kuang Repen, *Pierre* 4832 (lectotype FI-B!); Stung-Treng, *Choul* 2154 (K!). CHINA. Canton and Macao, Jan. 1837 (pist.), *Gaudichaud* 53 (P!); Simao mts, 1899 (stam.), *Henry* 12924 (CAL!, FI-B!, K!). Hainan: Na-ta, 1 Nov. 1921 (pist.), *McClure* 8038 (BM!, E!, K!, P!); Vo Lau to Na-ta, 1935 (pist.), *Linsley Gressit* 978 (BM!, E!). HONG KONG. Clearwater Bay, 31 May 1970 (stam., pist.), *Shiu Ying Hu* 10354 (K!); Tai Mo Shan, N.T., 150 m alt., 16 March 1995 (pist.), *Baker & Utteridge* BU9 (K!). MYANMAR. Pegu, 6 March 1871 (pist.), *Kurz* 3317 (BM!, CAL!); Maymyo plateau, 22 April 1913 (pist.), *Lace* 6165 (DD!, K!); Mindat, 16 March 1956 (stam.), *Kingdon-Ward* 21800 (BM!). PHILIPPINES. Batanes Is., Sabtang Island, June 1907 (pist.), *Fénix* (Bur. Sci.) 3744. (FI-B!); Mt Iraya, 11 May 1984, *Fernando* 403, 404 (K!). TAIWAN. Takao-san, 5 June 1912 (pist.), *Price* 585 (K!); Nan-Wan Heng-Chun, Pingtung County, 18 Dec. 1994 (stam., pist.), *Chu* 2 (K!). THAILAND. Kanchanaburi Prov., 10 km along road from Thing Pha Phum to Tripagodas, 8 Jan. 1994, *Barrow & Wongprasert* 7, 8, 9 (BKF!, K!); Chiang Mai Prov., Doi Inthanon, Vachiratharn Waterfall, 11 Jan. 1994, *Barrow & Phuma* 10, 11, 12 (BKF!, K!); Trang Prov., Thung Kai Forest Research Station, 22 Jan. 1994, *Barrow & Tingnga* 29, 30 (BKF!, K!). VIÊTNAM. Near Nha-Trang, 10 March 1922 (pist.), *Poilane* 2792 (P!); 25 June 1923 (ster.), *Poilane* 5932 (P!); Da Nang (Tourane), May – July 1927 (ster.), *Clemens* 4348 (K!); Central Annam, 25 Jan. 1930 (ster.), *Magalon* s.n. (P!).

VERNACULAR NAMES. Thailand. Peng [e.g., *Barrow* 1 (K)]. Philippines. Voyavoy (*Phoenix* palm), suot, vakol (coat made from leaflets of *Phoenix*) (Ivatan), [Greuzo & Fernando (1985)].

NOTES. *Phoenix* was first recorded in Indochina as *P. pusilla* by Loureiro (1790), who described a short palm from Hue in Vietnam. Unfortunately, the name *P. pusilla* had previously been used by Gaertner (1788) to describe a small palm of Sri Lanka and therefore *P. pusilla* Lour. is illegitimate. Kunth (1841) later described a palm from Vietnam as *P. loureiri* Kunth. The similarity between *P. loureiri* of Indochina and *P. humilis* of India was acknowledged by Beccari (1890), who included the former as one of five varieties of the latter in his monograph of the genus. This treatment was adopted by Beccari & Hooker (1892 – 93) and Blatter (1926). The name *P. humilis* (Royle 1840) was taken as the specific epithet, rather than *P. loureiri*, due to its earlier appearence in the literature, although it was not formerly described until 1890. The International Code of Botanical Nomenclature (Greuter *et al.* 1994) states that it is the date of description of a name that determines validity, and therefore *P. loureiri* (1841) must take precedence over *P. humilis* Royle ex Becc. (1890). This decision was reached by Moore (1963a) who accepted *P. loureiri* as a wide-ranging species including all varieties of *P. humilis* as defined by Beccari (1890, 1908). The type variety includes those *Phoenix* palms of Indochina and the Far East formerly referred to by Beccari (1890, 1908) as *P. humilis* var. *loureiri* and var. *hanceana*.

var. **humilis** *(Becc.) S. Barrow* **comb. nov.**, Royle, Ill. Bot. Himal. Mts. 1: 394, 397, 399, nom. (1840); Becc., Malesia 3: 373, t. 44, f. 2, 13 – 27 (1890); Becc. & Hook. f., Fl. Brit. India 6: 426 (1892); Osmaston, Forest Fl. Kumaon: 545 (1927); B. D. Naithani, Fl. Chamoli 2: 668 (1985). Type: NW India, (pist.), *Royle* s.n. (LE, photograph at K!).

P. ouseleyana Griff., Calcutta. J. Nat. Hist. 5: 347 (1845). No type designated. Griffith described the species from specimens communicated to him by Major Jenkins.

P. pedunculata Griff., Palms Brit. E. Ind.: 139 (1850). Type: Neilgherri Hill, June 1849, *Wight* 2767 (K!).

P. humilis Becc. var. *typica* Becc., Malesia 3: 379, 380, t. 44, 2, f. 22 – 24 (1890). Type: India, no precise locality, (pist.), *Royle* s.n. (photograph of specimen at K!).

P. humilis var. *pedunculata* (Griff.) Becc., Malesia 3: 379, 387, t. 44, f. 13 – 15, 18 – 21, 25 – 27 (1890). Type: Neilgherri Hill, June 1849, *Wight* 2767 (K!).

P. humilis var. *robusta* Becc., Malesia 3: 379, 384 (1890). Type: Bihar, Parasnath Hill, 1275 m alt., April 1884 (ster.), *Hook. f.* 643 (K!).

P. robusta Becc. & Hook. f., Fl. Brit. India 6: 427 (1892). Type: Bihar, Parasnath Hill, 1275 m alt., April 1884 (ster.), *Hook. f.* 643 (K!).

Distinguished from var. *loureiri* by the absence of sclerotic, tannin-filled cells in the leaflet margin and abaxial midrib region.

SELECTED SPECIMENS EXAMINED. BANGLADESH. Dacca, 15 April 1868 (stam., pist.), *Clarke* 6832 (FI-B!). INDIA: ANDHRA PRADESH. Ganjam Distr., March 1886, *Gamble* 17058 (K!). ASSAM. Jaintia Hills, 1350 m alt., Nurting, 16 Oct. 1867 (ster.), *Clarke* 5334 (FI-B!); Naga Hills, East of Kohima, 5 March 1935 (pist.), *Kingdon-Ward* 11131 (BM!). BIHAR. Parasnath Hill, Harazibagh, 10 Oct. 1883 (pist.), *Clarke* 33791 (FI-B!). HIMACHAL PRADESH. Una Distr., Mandi, 28 July 1977 (pist.), *Uniyal* 61254, 61255 (BSD!). KARNATAKA. Yellapore, 26 March 1905 (pist.), *Talbot* 4422 (CAL!, K!). KASHMIR. Painal, 9 Sept. 1986 (stam.), *Hajra* 82338 (BSD!). KERALA. Trivandrum, Ponmudi Hill, 940 m alt., 24 Jan. 1995 (pist.), *Barrow & Bunoi* 46, 47 (K!, TBGRI!). MADHYA PRADESH. Allahabad, Durgapur, Malandighi, 25 July 1973 (pist.), *Mukherjee* 18789 (CAL!). MAHARASHTRA. Nandgaon, April 1895 (stam., pist.), *Woodrow* s.n. (K!), 21 Feb. 1896 (stam.). ORISSA. Dandakaronya, Balimela, 21 May 1958 (pist.), *Rao* 18478 (CAL!). SIKKIM. Badawtaw, Nov. 1879 (stam.), *Gamble* 7340 (K!). TAMIL NADU. (Nilgiri Hills): Neilgherri Hill, June 1849 (pist.), *Wight* 2767 (BM!, CAL!, FI-B!, K!). (Coimbatore Distr.): Gudde Madappa, nr. Kotadi, 27 April 1934 (pist.), *Nagarathai* 81797 (K!). (Dindigul Distr.): Kodaikanal, rd. to Palani, 28 Jan. 1995, *Barrow & Kumar* 54, 55 (K!, RHT!). (Kuwool Distr.): Janpore, 28 July 1932, *Jacob* s.n. (K!). (South Arcot Distr.): North Arcot, Athanur-Yelagiri Hills, 18 March 1978, *Vajravelu* 53472 (CAL!). UTTAR PRADESH. Ranipur, 29 Jan. 1899 (pist.), *Kanjilal* 821 (K!); Gharwal Div., 27 May 1902 (pist.), *Duthie* 26021 (DD!, K!); Saharasdhara, 9 April 1963 (pist.), *Mahotra* 27122 (BSD!); Siwalik Hills, Rajaji National Park, c. 300 m alt., 26 Jan. 1995 (stam.), *Barrow* 44, 45 (K!, DD!). WEST BENGAL. Darjeeling, between Barnesbet and Singla Bazaar, 27°06'N, 88°17'E, 7 Aug. 1992 (pist.), *Long et al.* 1232 (E!). NEPAL. Seti Zone, Kailali Distr., 6 Dec. 1966 (pist.), *Nicholson* 2843, 2857 (BM!).

VERNACULAR NAMES. INDIA. Khajur, chota khajur, khajuri (Hindi); thakal (Kumaon region of Uttar Pradesh), [Blatter (1926)]. NEPAL. Thakal (Nepali), [Thumsi Hills, Nawal Parasi, *Makin* 1 (BM!), Rupandehi, near Khasyauli, *Makin* 233 (BM!)].

NOTES. The first reference to palms attributable to *P. loureiri* in India was made by Royle (1840), who noted the species (referred to as *P. humilis*) 'to be common in the Kheree Pass at 2,500 feet ..with *Pinus longifolia*.' Strangely, no mention was made of the species by Roxburgh (1832), despite its common occurrence throughout the subcontinent. Griffith (1845) described a new species, *P. ouseleyana* Griff., from Assam and Chota Nagpur in Bihar, which is referable to *P. humilis*, but he made no reference to the latter name. *Phoenix humilis* was formally described by Beccari (1890), as a variable and widespread species, comprising five varieties distributed in India, Indochina and the Far East. The three Indian varieties, *P. humilis* var. *typica*, var. *pedunculata* and var. *robusta*, are included here as *P. loureiri* var. *humilis* (Becc.) S. Barrow. The varieties of *P. humilis* as defined by Beccari (1890) do appear to reflect regional variation, but there are inadequate data to support their maintenance. *Phoenix humilis* var. *typica* Becc. refers to palms of northern India. *Phoenix humilis* var. *pedunculata* Becc. refers to a taxon of southern India which bears an infructescence peduncle that is very long relative to its short stem height. *Phoenix humilis* var. *robusta* Becc. refers to a poorly known robust taxon from the Pune region in Maharashtra, Parasnath Hill in Bihar, and a number of localities in Andhra Pradesh.

IMPERFECTLY KNOWN TAXON

Phoenix atlantica A. Chev., Bull. Mus. Hist. Nat. (Paris), sér. 2, 7: 137 (1935). Type: Isle de Sal, Algodeiro, 1934 (stam., pist.), *Chevalier* 45839 (P!).

SELECTED SPECIMENS EXAMINED. CAPE VERDE IS. Isle de Sal, Palmeira, Palha Verde (ster.), *Chevalier* 45840 (P!); San Thiago Praia, (pist.), *Chevalier* 45851 (P!); Isle de San Thiago, Praia, Sao Martinho, *Chevalier* 45854, 45858 (P!); Is. de Sal, Algodeiro, 1934 (stam., pist.), *Chevalier* 45839 (P!); Is. de Sal, Palmeira, Palha Verde, 1934 (ster.), *Chevalier* 45840 (P!); Is. de Sal, Pedro Lime, 26 June 1934 (stam.), *Chevalier* s.n. (P!).

NOTES. *Phoenix atlantica* was described by Chevalier (1952) from the Cape Verde Islands as a clustering species occurring in large clumps of 2 – 6 stems, to 15 m tall and 45 cm in diameter. Chevalier (1952) considered the taxon to have characteristics of *P. dactylifera*, *P. canariensis* and *P. reclinata*. All available evidence suggests to me that *P. atlantica* is a close relative of *P. dactylifera*, but, it is not yet clear whether the taxon merely represents feral date palms, or products of a series of hybridisation events between *P. dactylifera* and other species in the genus, or whether it is a distinct species. Further information is required before the species status of *P. atlantica* can be clarified.

Chevalier (1952) described *P. atlantica* var. *maroccana* A. Chev. from the Atlas Mountains of southwestern Morocco and reported large groves of 100,000 – 150,000 palms near Marrakech. The edible fruits were said to be sold locally in the markets but not exported. I consider this varietal name to refer to *P. dactylifera* and have included it here as a synonym of that name.

Excluded Names and Nomina Nuda

The following list of invalid and unpublished names includes those of horticultural origin, those associated only with herbarium specimens and those of uncertain application. A note is made where possible to the species to which an invalid name refers.

Fulchironia senegalensis Lesch., Desf. Cat. Pl., (ed. 3): 29, (1829), *nom. illeg.*, sine descr. = *P. reclinata* Jacq.

Palma vinifera Thevet, J. Bauh. Hist. 1, cap 160, 369.

Phoenicoidea Griff., J. Trav.: 46, *nom.* (1847) = *P. rupicola* T. Anderson.

Phoenix acaulis (auct. non Roxb.), Benth., Fl. Hongk.: 340 (1861) = *P. loureiri* var. *loureiri* Kunth.

P. andamanensis Hort. ex L. H. Bailey, Stand. Cycl. Hort.: 2594, *nom.* (1916) = *P. andamanensis* S. Barrow.

P. aequinoctialis Hort., Rev. Hort.: 340 (1879). A name of unknown application and no botanical standing.

P. aequinoctialis leonensis Hort., Rev. Hort.: 340 (1879). A name of unknown application and no botanical standing.

P. aequinoctialis spinosa Hort. Rev. Hort.: 340 (1879). A name of unknown application and no botanical standing.

P. ammonis Ehrenb. ex Steud, Nomencl. Bot. (ed. 2) 2: 323, *nom.* (1841). A name of unknown application and with no botanical standing.

P. andersonii Hort. ex Gentil, Pl. Cult. Serres, Jard. Bot. Brux. 148, *nom.* (1907); Cat. Hort. Calc. 119 (1886 – 7); Gard. Chron. 2: 45 (1887) = *P. rupicola* T. Anderson.

P. atlantidis A. Chev., Compt. Rend. Hebd. Séances Acad. Sci. 199: 1153, *nom.* (1934) = *P. atlantica* A. Chev.

P. butia Usteri, Guia Bot. Praca Rep. Jard. Luz.: 13, *nom.* (1919). A name of unknown application and with no botanical standing.

P. canariensis cycadifolia Hort. ex A. Chev., Rev. Int. Bot. Appl. Agric. Trop.: 32: 220 (1952). *nom nud.* Chevalier (1952) notes this cultivated variety to have 'more curved leaflets than the type' = *P. canariensis* Chabaud.

P. capoverdensis A. Chev., *nom. in sched.*, Isle de Sal, Cape Verde Is., 24 – 30 June 1934 (stam.), *Chevalier* 45850 (P!) = *P. atlantica* A. Chev.

P. cycadifolia macrocarpa Hort., Bonnel in Rev. Hort.: 170, *nom.* (1881) = *P. canariensis* Chabaud.

P. dactylifera var. *canariensis* Hort. ex Regel, Gartenflora 28: 131 (1879); ex Drude, Gartenzeitung 1: 182, fig.182 (1882). Drude (1882) considered *P. canariensis*, *P. cycadifolia* and *P. tenuis* to be merely forms of *P. dactylifera*. Drude (1882) illustrates *P. dactylifera* var. *canariensis*, but gives no description = *P. canariensis* Chabaud.

P. dactylifera var. *excelsa* Hort. ex L. H. Bailey, Stand. Cycl. Hort.: 2594, *nom.* (1916) = *P. dactylifera* L.

P. dumosa Hort. ex L. H. Bailey, Stand. Cycl. Hort.: 2594, *nom.* (1916). A name of unknown application and with no botanical standing.

P. excelsa Steud. (sphlam. = *excelsior*), Nomencl. Bot. ed. 2: 323, *nom.* (1841) = *P. dactylifera* L.

P. equinoxialis Bojer, Hortus Maurit.: 306, *nom.* (1837) = *P. reclinata* Jacq.

P. farinifera (auct. non Roxb.), Hance, J. Bot. 7: 15 (1869); Benth., Fl. Hongk. Suppl.: 41 and J. Linn. Soc., Bot. 13: 129, *nom.* (1873) = *P. loureiri* var. *loureiri* Kunth.

P. ferruginea Hort. ex H. Wendl., Index Palm: 31 (1854), *nom.* A name of unknown application and with no botanical standing.

P. gibsoni, nom. in sched. A name associated only with an unidentified specimen of *Phoenix* in the Kew herbarium (India, Poona, 20 Jan. 1909 (ster.), *Herbarium on Economic Products* 30686-7).

P. gracillima Hort. A name of unknown application and no botanical standing. No published reference has been found.

P. hanceana Naudin ex. Hance, J. Bot. 27: 174, *nom.* (1879) = *P. loureiri* var. *loureiri* Kunth.

P. jubae var. *edulis* A. Chev., Rev. Int. Bot. Appl. Agric. Trop. 4: 196, *nom. inval.* (1924). No description = *P. canariensis* Chabaud.

P. leonensis Lodd., Palms: 9, *nom.* (1845); Kunth, Enum. Pl. 3: 256, *nom.* (1841) = *P. reclinata* Jacq.

P. lilliput Hort., Rev. Hort.: 340, *nom.* (1879). A name of unknown application and with no botanical standing.

P. medinensis, nom. in sched. Noted on a specimen in the Edinburgh herbarium (*Chaudhary* E421) of *P. reclinata* from Saudi Arabia = *P. reclinata* Jacq.

P. natalensis Hort. ex L. H. Bailey, Stand. Cycl. Hort.: 2594, *nom.* (1916) = *P. reclinata* Jacq.

P. paradenia Hort. ex L. H. Bailey, Stand. Cycl. Hort.: 2594, *nom.* (1916). A name of unknown application and with no botanical standing.

P. pumila Regel, Gartenflora 20: 153 (1871), *nom. nud.* A name used to refer to an unidentified *Phoenix* taxon (probably *P. reclinata*) from Gabon.

P. pusilla var. *andamanensis* Becc., *nom. in sched.* Beccari named three specimens (*Rogers* s.n., 132 and 285) in the Kew herbarium with this name, but no published reference or description exists = *P. andamanensis* S. Barrow.

P. pygmaea Raeusch., Nomencl. Bot. (ed. 3): 375, *nom.* (1797) = *P. loureiri* var. *loureiri* Kunth.

P. pygmaea Hort. ex Link, Enum. Hort. Berol. Alt. 2: 421, *nom.* (1822). A name of unknown application and no botanical standing.

P. spadicea Wight ex Mart., Hist. Nat. Palm. 3 (ed. 2): 274 (1849). An unpublished name associated by Martius (1849) with *P. farinifera* Roxb.

P. subhypogaea C. B. Clarke, *nom. in sched.* A name found only on a collection from Uttar Pradesh in India (between Saharanapur and Najiiabad in Bijnor, 10 March 1887 (pist.), *Duthie* s.n. (DD, K)) = *P. acaulis* Roxb.

P. tenuis Hort. Verschaff., Cat. 1863: 13 cum. ic., ex Neubert, Deutsch. Mag. Garten-Blumenk. 26: 203 – 205, fig. 204 (1873); Sauvaigo in Rev. Hort. 66: 495 (1894). *P. tenuis* is an early name for the Canary Island palm, though no description accompanies it. Neubert (1873) considered *P. tenuis* and *P. canariensis* to be very close. According to *Index Londinensis*, *P. tenuis* was illustrated as early as 1863 by Verschaffelt, Catalogue (1863): 13 cum ic. = *P. canariensis* Chabaud.

Phoenix tillaensis Becc., *nom. in sched.* This name is found in connection with specimens of one collection (India, Punjab, Salt range, Mt Tilla, May 1890, *McDonnell* s.n.). A name of unknown application and no botanical standing = *P. sylvestris* Roxb.

P. tomentosa Hort. ex Gentil., Pl. Cult. Serres Jard. Bot. Brux. 149 (1907). A name of unknown application and with no botanical standing.

P. vigieri Hort. ex Naudin, Rev. Hort. 57: 54 (1885) = *P. canariensis* Chabaud.

P. weddeliana Usteri, Guia Bot. Praca Rep. Jard. Luz.: 13, *nom.* (1919). A name of unknown application and with no botanical standing.

P. woodrowii, nom. in sched. A name noted on specimens in several herbaria (BM, DD, K) from Nandgaon in the Western Ghats of India [April 1895, *Woodrow* s.n. (K!); 1897, *Woodrow* s.n. (BM!, DD!)]. The name refers to a robust form of *P. loureiri* var. *humilis*, formerly referred to as *P. robusta* by Beccari & Hooker (1892 – 93).

P. zanzibarensis Hort. ex Gentil., Pl. Cult. Serres Bot. Brux.: 149, *nom.* (1907) = *P. reclinata* Jacq.

NAME WRONGLY ASSIGNED TO *Phoenix*.

P. humilis Cav., Icon. 2: 12, t. 115 (1793). The name *P. humilis* was used by Cavanilles to refer to *Chamaerops humilis* L. The flowers and fruit illustrated in Tab. 115 are clearly referable to *C. humilis* L.

INVALID NAMES REFERRING TO HYBRIDS

The existence of hybrids between *P. dactylifera* and other species of the genus, particularly *P. canariensis*, is well documented (e.g., Chevalier 1952; Moore 1971a; Naudin 1893; Sauvaigo 1894). The following names are invalid and refer to putative hybrids of *Phoenix*.

Microphoenix decipiens Naudin[2].

M. sahuti Carrière, Rev. Hort. 57: 513 – 514 (1885)[2].

P. canariensis var. *edulis* R. Prosch. ex A. Chev., Rev. Int. Bot. Appl. Agric. Trop. 1: 223 (1921), *nom. nud.*[3]

P. canariensis var. *erecta* Hort. ex A. Chev., Rev. Int. Bot. Appl. Agric. Trop. 32: 220 (1952), *nom. nud.*[3]

P. canariensis var. *glauca* Hort. ex A. Chev., Rev. Int. Bot. Appl. Agric. Trop. 32: 220 (1952), *nom. nud.*[3]

P. canariensis var. *macrocarpa* Hort. ex A. Chev., Rev. Int. Bot. Appl. Agric. Trop. 32: 220 (1952), *nom. nud.*[3]

P. canariensis var. *tenuis* Hort., *nom. nud.*[3]

---

[2] Species of *Phoenix* are well known for their ability to cross with each other producing a range of a hybrid taxa. Reports of hybrids between *P. dactylifera* and other genera of palms also exist. Naudin described *Microphoenix decipiens* to refer to a supposed hybrid between *Chamaerops humilis* and *P. dactylifera*. Carrière (1885) described *M. sahuti* to refer to a hybrid taxon resulting from a cross between *M. decipiens* and *Trachycarpus excelsus* H. Wendl. Anatomical characteristics of *Microphoenix* and its reported parents were studied by Bargagli-Petrucci (1900) who considered the fruits of *M. decipiens* to be anatomically identical with those of *Chamaerops*. Carpological specimens from a cultivated palm grown in the garden of Professor Henriques in 1889, annotated as *Microphoenix decipiens*, are held in the Central Herbarium in Florence. The fruits are larger and more elongate than is typical of *Chamaerops*, thus appearing date-like. However, in my opinion these are not the fruit of a *Chamaerops* × *Phoenix* hybrid but of a large-fruited form of *Chamaerops*.

[3] One of several names referring to '*canariensis*-like' palms in horticulture (see Moore 1971a). These names could refer to hybrid taxa.

P. erecta Hort. ex Sauvaigo, Rev. Hort. 66: 495 (1894), *nom. nud.*[4]

P. hybrida Hort., Cat. Pl. Hort. Bogor: 72 (1866), *nom. nud.* The name *P. hybrida* was noted by Popenoe (1973) to refer to a hybrid between *P. dactylifera* and *P. canariensis.*

P. intermedia Hort., Becc., Malesia 3: 364 (1890), *nom. nud.* Beccari (1890) noted this name to refer to a hybrid between *P. dactylifera* and *P. canariensis.*

P. macrocarpa Hort., Becc., Malesia 3: 364 (1890), *nom. nud.* Beccari (1890) noted this name to refer to a hybrid between *P. dactylifera* and *P. canariensis*[4].

P. marioposae Hort., Sauvaigo, Bull. Soc. Agric. Alpes marit. (1891), *nom. nud.* This name is of unknown application but Chevalier (1952) suggests it refers to a hybrid between *P. dactylifera* and *P. spinosa* or *P. senegalensis.*

P. melanocarpa Naudin, Rev. Hort.: 563, fig. 178 (1893), Rev. Hort.: 494, fig. 182 – 4 (1894), *nom. nud.* Naudin (1893) regarded this name as referring to a hybrid between *P. sylvestris* and *P. canariensis.* The palm to which it refers bears black, seedless fruits from unpollinated flowers. Dr Proschowsky of Nice thought it merely a date palm. Naudin (1893) discusses the origin of a black-fruited date palm growing in Nice and Sauvaigo (1894) mentions such a palm in 'Les *Phoenix* Cultivées dans les Jardins de Nice'. Two hypotheses are proposed by these authors. Firstly, that *P. melanocarpa* is simply a variety of *P. dactylifera.* Alternatively, *P. melanocarpa* is proposed as a hybrid between *P. dactylifera* and either *P. canariensis* or *P. senegalensis* (which has small black fruit).

P. senegalensis Van Houtte ex Salomon, Gartenflora: 305 (1882), *nom. nud.* No reference has been found in the cited issue of Gartenflora. The application of this invalid name is confused.

Acknowledgements

I am grateful to The Royal Botanic Gardens, Kew, for financial support and access to all facilities. I owe much to John Dransfield for support and encouragement, and am very grateful to him, Chris Humphries and Barbara Pickersgill for their supervision. I thank Mark Chase and Tim Lawrence for their guidance in the molecular and anatomy laboratories. I am grateful to the curators of the following herbaria for access to material for study: BK, BKF, BM, BSD, CAL, DD, E, FI, FII, ISTO, K, P, PDA, PSI, RHT. For assistance in the field I am in debt to many people but in particular to Rachan Phuma of Huay Kaew Arboretum in Chiang Mai and the staff of the Royal Forest Department in Bangkok, to Professor Melih Boydak of the University of Istanbul, to Father K. M. Matthew at the Rapinat Herbarium in Tiruchirapolli, to the staff of the Tropical Botanic Garden and Research Institute in Trivandrum, and to Aruna Weerasooriya and the staff of the National Herbarium in Peradeniya, Sri Lanka. Scott Zona, of Fairchild Tropical Garden in Miami, acted as referee for this paper, and many thanks are given to him.

---

[4] Sauvaigo (1894) notes that the horticultural industry uses the name *P. tenuis* Hort. Verschaff. to refer to young individuals of *P. canariensis,* and the species has given rise to several varieties including *P. macrocarpa* and *P. erecta* as a result of hybridisation with *P. dactylifera, P. senegalensis* and *P. reclinata.*

REFERENCES

Aitchson, J. E. T. (1869). A Catalogue of the Plants of the Punjab and Sindh. Taylor & Francis, London.

Alvin, K. L. & Boulter, F. L. S. (1974). A controlled method of comparative study for Taxodiaceous leaf cuticles. Bot. J. Linn. Soc. 69: 277 – 286.

Barclay, C. (1974). A new locality of wild *Phoenix* in Crete. Ann. Mus. Goulandris 2: 23 – 29.

Barfod, A. (1988). Leaf anatomy and its taxonomic significance in phytelephantoid palms (*Arecaceae*). Nordic J. Bot. 8 (4): 341 – 348.

Bargagli-Petrucci, G. (1900). Riceche anatomiche sopra la *Chamaerops humilis* L., la *Phoenix dactylifera* L. ed i loro prestesi ibridi (*Microphoenix*). Malpighia 14: 306 – 360, t. 8 – 13.

Barrow, S. (1994). In search of *Phoenix roebelenii*: the Xishuangbanna palm. Principes 38 (4): 177 – 181.

Baum, B. R. & Appels, R. (1992). Evolutionary change at the 5S DNA loci of species in the *Triticeae*. Pl. Syst. Evol. 183: 195 – 208.

—— & Donoghue, M. J. (1995). Choosing among alternative "phylogenetic" species concepts. Syst. Bot. 20 (4): 560 – 573.

Bavappa, K. V. A. & Nair, M. K. (1982). Cytogenetics and breeding. In: K. V. A. Bavappa *et al.* (eds.). The areca nut palm, pp. 51 – 96. Central Plantation Crops Research Institute, Kasaragod, India.

Beal, J. M. (1937). Cytological studies in the genus *Phoenix*. Bot. Gaz. 99: 400 – 407.

Beccari, O. (1890). Revista monografica delle species del genera *Phoenix* L. Malesia 3: 345 – 416.

—— (1906). Palmarum Madagascariensium Synopsis. Bot. Jahrb. Syst. 38 (87): 4.

—— (1908). The palms of the Batanes and Babuyanes Islands. Philipp. J. Sci. 3 (6): 339 – 340.

—— (1914). Palme del Madagascar. Istituto Micrografica Italiano, Firenze.

—— & Hooker, J. D. (1892 – 93). *Palmae*. In: J. D. Hooker, The Flora of British India 6: 402 – 483. L. Reeve, London.

Beentje, H. J. (1994). Kenya Trees, Shrubs and Lianas: 644. National Museums of Kenya, Nairobi.

Berry, E. W. (1914). Fruits of a date palm in the Tertiary deposits of eastern Texas. Amer. J. Sci. 187: 403 – 406.

Blatter, E. (1926). The Palms of British India and Ceylon. Oxford University Press, Mangalore, India.

Boissier, P. E. (1882). Flora Orientalis 5: 47. H. Georg, Basel.

Bonavia, E. (1885). The Date Palm. Gard. Chron. 24: 178 – 211.

Boydak, M. (1983). Ülkemizin nadide bir dogal türü Datça Hurmasi (*Phoenix theophrasti* Greuter). Çevre Koruma 18: 20 – 21. Istanbul.

—— (1985). The distribution of *Phoenix theophrasti* in the Datça Peninsula, Turkey. Biol. Conservation 32: 129 – 135.

—— (1986). Kumlaca-Karaöz'de Saptanan Yeni bir Dogal Palmiye (*Phoenix theophrasti*) yayilisi. Istanbul Üniv. Orman Fak. Derg., A 36-1: 1 – 13.

—— (1987). A new occurrence of *Phoenix theophrasti* in Kumlaca-Karaöz, Turkey.

Principes 31 (2): 89 – 95.

—— & Barrow, S. (1995). A new locality for *Phoenix* in Turkey: Gölköy-Bödrum. Principes 39 (3): 117 – 122.

—— & Yaka, M. (1983). Datça Hurmasi (*Phoenix theophrasti*) ve Datça Yarimadasinda saptanan dogal yayilisi. Istanbul Üniv. Orman Fak. Derg., A 33-1: 73 – 92.

Brac de la Perrière, R. A. (1988). Recherche sur les resources génétiques sur palmier dattier en Algérie. Ann. Inst. Natl. Agron. (El-Harrach) 12 (1): 493 – 506.

Brandis, D. (1906). Indian Trees: 646. A. Constable & Co., London.

Burret, M. (1943). Die Palmen Arabiens. Bot. Jahrb. Syst. 73 (2): 188 – 190.

Buzek, C. (1977). Date palm seeds from the lower Miocene of Central Europe. Vestn. Ustredn. Geol. 52: 159 – 168.

Carrière, E. A. (1885). *Microphoenix sahuti.* Rev. Hort. 57: 513 – 514.

Chabaud, B. (1882). La *Phoenix canariensis.* La Provence Agricole et Horticole Illustrée 19: 293 – 297, fig. 66 – 68.

Chandler, M. E. J. (1961 – 64). The Lower Tertiary Floras of Southern England. 4 vols. British Museum (Natural History), London.

Chevalier, A. (1952). Recherches sur les *Phoenix* africains. Rev. Int. Bot. Appl. Agric. Trop. 32: 205 – 236, 355 – 356.

Chiovenda, E. (1929). Flora Somala 1: 317 – 318. Sindacato Italiano, Roma.

Christ, D. H. (1885). Vegetation and Flora der Canarischen Inseln. Bot. Jahrb. Syst. 6: 469.

Collenette, S. (1985). An Illustrated Guide to the Flowers of Saudi Arabia. Scorpion Publishing Ltd., London.

Conwentz, H. (1886). Die Flora des Bernsteins II. W. Engelmann, Leipzig, Germany.

Corner, E. J. H. (1966). The Natural History of Palms. University of California Press, Berkeley.

Cowan, P. J. (1984). Impaled Dates: By shrikes or by chance? Bull. Ornithological Soc. Middle E. 13: 7 – 8.

Cox, A. V., Bennett, M. D. & Dyer, T. A. (1992). Use of the polymerase chain reaction to detect spacer size heterogeneity in plant 5S rRNA gene clusters and to locate such clusters in wheat (*Triticum aestivum* L.). Theor. Appl. Genet. 83: 684 – 690.

Daghlian, C. P. (1978). Coryphoid palms from the Lower and Middle Eocene of southeastern North America. Palaeotonographica Abt. B 166: 44 – 82.

Davis, T. A. (1972). Tapping the wild date. Principes 16 (1) : 12 – 15.

De Candolle, A. (1884). The Origin of Cultivated Plants. Trench & Company, London.

DeMason, D. A., Stolte, K & Tisserat, B. (1982). Floral development in *Phoenix dactylifera* L. Canad. J. Bot. 60 (8): 1437 – 1446.

—— & Tisserat, B. (1980). The occurrence and structure of apparently bisexual flowers in the date palm, *Phoenix dactylifera* L. (*Arecaceae*). Bot. J. Linn. Soc. 181: 283 – 292.

De Zoysa, N. (in press). *Palmae.* In: Flora of Ceylon. Amerind Publishing Co., New Dehli.

Dhar, Shri (1998). *Phoenix acaulis.* Principes 42 (1): 11 – 12.

Doyle, J. J. & Doyle, J. L. (1987). A rapid DNA isolation procedure for small quantities of fresh leaf tissue. Phytochem. Bull. 19: 11 – 15.

Dransfield, J. (1985). Flora of Iraq 147 (*Palmae*): 262 – 265. Ministry of Agriculture, Baghdad.

—— (1992). Observations on rheophytic palms in Borneo. Bull. Inst. Franç. Études Andines 21 (2): 415 – 432.

—— & Beentje, H. J. (1995). *Satranala* (*Coryphoideae*: *Borasseae*: *Hyphaeninae*), a new palm genus from Madagascar. Kew Bull. 50 (1): 85 – 92.

Drude, O. (1887). *Palmae*. In: A. Engler & K. Prantl (eds.), Die Natürlichen Pflanzenfamilien 2 (3), pp. 1 – 93. W. Engelmann, Leipzig.

Eggeling, W. J. (1940). The Indigenous Trees of the Uganda Protectorate: 164. Government printer, Entebbe, Uganda.

Fischer, T. (1882). The Date Palm. Petermann's Mittheilungen: 64 (1881). A review in Bot. Jahrb. Syst. 2: 230 (1882) & Bull. Soc. Bot. France 29: 110 (1882).

Gaertner, J. (1788). De Fructibus et Seminibus plantarum 1: 24, tab. 9. Academiae Carolinae, Stuttgart.

Gamble, J. S. (1902). A Manual of Indian Timbers. Office of the Superintendent of Government Printing, Calcutta, India.

Gibbon, E. (1776 – 1788). Decline and Fall of the Roman Empire.

Goor, A. (1967). The history of the date through the ages in the Holy Land. J. Econ. Taxon. Bot. 21: 320 – 340.

Greuter, W. (1967). Beiträge zur Flora der Südägäis 8 – 9. Bauhinia 3 (2): 243 – 250.

——, Barrie, F. R., Burdet, H. M., Chaloner, W. G., Demoulin, V., Hawksworth, D. L., Jørgenson, P. M., Nicholson, D. H., Silva, P. C., Trehane, P. & McNeill, J. (eds.). (1994). International Code of Botanical Nomenclature (Tokyo Code). Koeltz Scientific Books, Königstein, Germany.

Griffith, W. (1845). Palms of British East India. Calcutta J. Nat. Hist. 5: 1 – 103.

Grisebach, A. H. R. (1872). Die Vegetation der Erde. W. Engelmann, Leipzig.

Gruezo, W. S. & Fernando, E. S. (1985). Notes on *Phoenix hanceana* var. *philippinensis* in the Batanes Islands, Philippines. Principes 29 (4): 170 – 176.

Hamilton, A. C. (1981). A field guide to Ugandan Forest Trees. Makerere University, Uganda.

Hamilton, F. (1827). A commentary on the Hortus Malabaricus. Trans. Linn. Soc. London 15 (1): 83.

Hehn, V. (1888). The Date Palm. In: V. Hehn (ed.), The Wanderings of Plants and Animals from their first Home, p. 203. S. Sonnenschein & Co., London.

Henderson, A. (1986). A review of pollination studies in the *Palmae*. Bot. Rev. (Lancaster) 52: 221 – 259.

Hermann, P. (1687). Horti Academici Lugduno-Batavi Catalogus. C. Boutesteyn, Leiden.

—— (1698). Paradisus Batavus. A. Elzevier, Leiden.

—— (1717). Musaeum Zeylanicum. I. Severinus, Leiden.

Herrera, J. (1989). On the reproductive biology of the Dwarf Palm, *Chamaerops humilis* L. in Southern Spain. Principes 33 (1): 27 – 32.

Higgins, D. G., Fuchs, R. & Blesby, A. (1992). CLUSTAL: a new multiple sequence alignment program. Computer Applic. Biosci. 8: 189 – 191.

Hodel, D. R. (1995). *Phoenix*: The Date Palms. Palm J. 122: 14 – 36.

Hort, Sir A. (1916). Theophrastus' Enquiry into Plants and minor works on odours and weather signs bk. 6, cap. 37. With an English translation by Sir Arthur Hort, Heinemann, London.

Jumelle, H. & Perrier de la Bâthie, H. (1913). Palmiers de Madagascar. Ann. Inst. Bot.-Géol. Colon. Marseille sér. 3, 1: 1 – 91.

—— & —— (1945). Palmiers. In: H. Humbert (ed.) Flore de Madagascar et des Comores. Imprimerie officielle, Antananarivo.

Kaempfer, E. (1689). *Palmae* Persicae Historia, Bl. 26 – 33.

—— (1712). *Phoenix persicus*: Die Geschichte der Dattelpalmae. See Muntschick (1987).

Kaplan, D. R., Dengler, N. G. & Dengler, R. E. (1982). The mechanism of plication inception in palm leaves: problem and developmental morphology. Canad. J. Bot. 60: 2999 – 3106.

Kiew, R. (1988). Portraits of threatened plants 16. *Phoenix paludosa* Roxb. Malayan Naturalist 42 (1): 16.

Kinnaird, M. F. (1992). Competition for a forest palm: use of *Phoenix reclinata* Jacq. by human and non-human primates. Conservation Biol. 6 (1): 101 – 107.

Kunth, K. S. (1841). Enumeratio Plantarum 3: 257. J.G. Cottae, Stuttgart & Tübingen.

Kurz, W. S. (1870). Report on the Vegetation of the Andaman Islands. Office of the Superintendent of Government Printing, Calcutta, India.

—— (1877). Forest Flora of British Burma 2. Office of the Superintendent of Government Printing, Calcutta, India.

Linnaeus, C. (1736). Musa Cliffortiana. Leiden.

—— (1747). Flora Zeylanica. Stockholm.

—— (1753). Species Plantarum. 2 vols. Stockholm.

Loureiro, J. de. (1790). Flora cochinchinensis. Lisbon.

Lüpnitz, D. & Kretschmar, M. (1994). Standortsökologishce Untersuchungen an *Phoenix canariensis* Hort. ex Chabaud (*Arecaceae*) auf Gran Canaria und Teneriffa (Kanarische Inseln). Palmarum Hortus Francofurtensis 4: 23 – 63.

Machin, J. (1971). Plant microfossils from Tertiary deposits of the Isle of Wight. New Phytol. 70: 851 – 872.

Maddison, W. P. & Maddison, D. R. (1992). MacClade 3.01. Sinauer Associates Inc., Sunderland, Massachusetts, USA.

Magalon, M. (1930). Contribution a l'étude des palmiers de l'Indochine Française, pp.20 – 30. Theses Presentées á la Faculté des Sciences de Montpellier.

Mahabalé, T. S. & Parthasarathy, M. V. (1963). The genus *Phoenix* L. in India. J. Bombay Nat. Hist. Soc. 60 (2): 371 – 387.

Mai, P. H. & Walther, H. (1978). Die Floren der Haselbacher Serie im Weisselster-Becken (Leipzig). Abh. Staatl. Mus. Mineral. & Geol. Dresden 28: 1 – 200.

Malik, K. A. (1984). Flora of Pakistan 153: 20 – 25. Pakistan Agricultural Research Council, Islamabad.

Martens, J. & Uhl, N. W. (1980). Methods for the study of leaf anatomy of palms. Stain Technol. 55: 241 – 246.

Martius, C. F. P. von. (1823 – 1853). Historia Naturalis Palmarum. 3 vols. T. O. Weigel, Munich.

Mathew, S. P. & Abraham, S. (1994). The Vanishing Palms of the Andaman and Nicobar Islands, India. Principes 38 (2): 100 – 104.

Matthew, K. M. (1983). The Flora of the Tamilnadu Carnatic. Madras, India.

Mifsud, S. (1995). Palms on the Maltese Islands. Principes 39 (4): 190 – 196.

Miller, P. (1754). The Gardeners Dictionary (abr. ed. 4). S. Powell, Dublin.

Moore, H. E. (1963a). An annotated checklist of cultivated palms. Principes 7 (4): 119 – 184.

—— (1963b). Typification and species of Palma Miller. Gentes Herb. 9 (3): 235 – 244.

—— (1971a). *Phoenix canariensis* and *P. cycadifolia*. Principes 15 (1): 33 – 35.

—— (1971b). Wednesdays in Africa. Principes 15 (4): 111 – 119.

—— & Dransfield, J. (1979). The typification of Linnean palms. Taxon 28 (1, 2/3): 59 – 70.

Muntschick, W. (1987). Translation of '*Phoenix persicus*: The History of the Date Palm' by E. Kaempfer (1712). Basilisken-Presse, Marburg, Germany.

Naudin, C. V. (1893). Le dattier a fruits noirs. Rev. Hort. 65: 563 – 564.

Neubert, W. (1873). Über Palmen. Deutsch. Mag. Garten-Blumenk. 26: 203 – 205.

Nixon, R. W. (1951). The date palm 'Tree of Life' in subtropical deserts. Econ. Bot. 5: 274 – 301.

Noltie, H. J. (1994) Flora of Bhutan 3 (1): 234 – 236. Royal Botanic Garden, Edinburgh.

Padmanabhan, D. (1963). Leaf development in palms. Curr. Sci. 32: 537 – 539.

Padmanabhan, D. & Sudhersan, C. (1988). Mass destruction of *Phoenix loureiri* in South India. Principes 32(3): 18 – 123.

Parkinson, C. E. (1923). A Forest Flora of the Andaman Islands: 263. International Book Distributers, Dehra Dun, India.

Parrott, J. (1980). Frugivory by great grey shrikes *Lanius excubitor*. Ibis 122: 532 – 533.

Periasamy, K. (1967). Morphological and ontogenetic studies in palms. III. Growth patterns of the leaves of *Caryota* and *Phoenix* after the initiation of plications. Phytomorphology 16: 474 – 490.

Petter, J.-J., Albignac, R. & Rumpler, Y. (1977). Faune de Madagascar 44L mammifères lemuriens (Primates Prosimiens). ORSTOM/CNRS, Paris.

Popenoe, P. (1924). The date palm in antiquity. Sci. Monthly 19: 313 – 325.

—— (1973). The Date Palm. Field Research Projects, Coconut Grove, Miami.

Rackham, H. (1945). Pliny: Natural History XIII (6 – 9), translated by H. Rackham, Loeb Classical Library vol. 4. Cambridge, Massachusetts.

Rao, P. S. N. (1996). Phytogeography of the Andaman and Nicobar Islands, India. Malayan Nat. J. 50: 57 – 79.

Ray, J. (1686 – 1704). Historia Plantarum. 3 vols. London.

Read, R. W. & Hickey, L. J. (1972). A revised classification of fossil palm and palmlike leaves. Taxon 21: 129 – 137.

Regel, E. (1879). *Phoenix cycadifolia* h. Athen. Gartenflora 28: 131.

Reid, E. M. & Chandler, M. E. J. (1933). The London Clay flora. British Museum (Natural History), London.

Rheede, Tot Drakenstein, H. Van. (1678 – 1693). Hortus Indicus Malabaricus 3: 15 – 16, pl. 22 – 25. J. van Someren & J. van Dyck, Amsterdam.

Ridley, H. N. (1930). The dispersal of plants throughout the world. L. Reeve & Co., Ashford.

Roxburgh, W. (1832). Flora Indica 3: 783 – 790. Mission Press, Serampore, India.

Royle, J. F. (1840). Illustrations of the Botany of the Himalayan Mountains 1: 394, 397, 399. W. H. Alland & Co., London.

Saiki, R. K., Gelfand, S., Stoffel, S., Scharf, R. H., Hifuchi, G. K, Mullis, K. B. & Erlich, H. A. (1988). Primer-directed enzymatic amplification of DNA with a thermostable DNA polymerase. Science 239: 487 – 491.

Sanger, F., Nicklen, S. & Coulson, A. R. (1977). DNA sequencing with chain terminating inhibitors. Proc. Natl. Acad. Sci. U.S.A. 74: 5463 – 5467.

Sastri, D. C., Hilu, K., Appels, R., Lagudah, E. S., Playford, J. & Baum, B. (1992). An overview of evolution in plant 5S DNA. Pl. Syst. Evol. 183: 169 – 181.

Sauvaigo, E. (1894). Les Phoenix cultivés dans les jardins de Nice. Rev. Hort. 66: 493 – 499.

Schmidt, M. E. (1994). The distribution and characteristics of Louisiana petrified palmwood. Principes 38 (3): 142 – 145.

Schonland, S. (1924). On the theory of 'age and area'. Ann. Bot. (London) 38: 453 – 472.

Schweinfurth, G. (1873). The Heart of Africa 1: 127. Marston, Low & Searle, London.

Smitinand, T. (1948). Thai Plant Names. Royal Forest Department, Thailand.

Solecki, R. S. & Leroi-Gourhan, A. (1961). Palaeoclimatology and archaeology in the Near East. Ann. New York Acad. Sci.: 729 – 739.

Swofford, D. L. (1990). PAUP: Phylogenetic Analysis using Parsimony, Version 3.1.1. Computer program distributed by the Illinois Natural History Survey, Champaign, Illinois, U.S.A.

Täckholm, V. & Drar, M. (1950). Flora of Egypt 2: 165 – 273. Fouad I University Press, Cairo.

Theophrastus. (370 – 285 BC). Enquiry into Plants and minor works on odours and weather signs bk. 6, cap. 37. (see Hort 1916).

Thulin, M. (1995). Flora of Somalia 4: 271 – 272. Royal Botanic Gardens, Kew.

Thwaites, G. H. K. (1864). Enumeratio Plantarum Zeylaniae. Dulau & Co., London.

Tomlinson, P. B. (1961). Anatomy of the monocotyledons II. Palmae. Clarendon Press, Oxford.

Tomlinson, P. B. (1990). The Structural Biology of Palms. Clarendon Press, Oxford.

Trimen, H. (1885). Notes on the Flora of Ceylon. J. Bot. 23: 173, 267.

—— (1898). A Handbook to the Flora of Ceylon 4: 326 – 27, pl. 95. Dulau & Co., London.

Troupin, G. (1987). Flore de Rwanda 4: 399, f. 170. Tervuren, Belgium.

Tuley, P. (1995). The Palms of Africa. The Trendine Press, St. Ives.

Turland, N. J., Chilton, L. & Press, J. R. (1993). Flora of the Cretan area: An annotated checklist and atlas. HMSO, London.

Uhl, N. W. & Dransfield, J. (1987). Genera Palmarum: A classification of palms based on the work of Harold E. Moore, Jr. L. H. Bailey Hortorium and the International Palm Society, Lawrence, Kansas, USA.

—— & Moore, H.E. (1971). The palm gynoecium. Amer. J. Bot. 58: 945 – 992.

Webb, P. B. & Berthelot, S. (1847). Histoire Naturelle des Iles Canaries 3 (2): 289. Paris.

Werth, E. (1934). Zur Kultur der Dattelpalme und die Frage ihrer Herkunft. Ber. Deutsch. Bot. Ges. 51: 501 – 512.

Whitmore, T. C. (1973). Palms of Malaya. Oxford University Press, Kuala Lumpur.

Zohary, D. & Hopf, M. (1988). Domestication of plants in the Old World: the origin and spread of cultivated plants in West Asia, Europe and the Nile Valley. Clarendon Press, Oxford.

—— & Speigel-Roy, P. (1975). Beginnings of fruit growing in the Old World. Science 187: 319 – 327.

Zona, S. (1990). A monograph of *Sabal* (*Arecaceae: Coryphoideae*). Aliso 12(4): 583 – 666.

Zona, S. & Henderson, A. (1989). A review of animal-mediated seed dispersal in palms. Selbyana 11: 6 – 21.